PROBLEMS FOR MATHEMATICIANS, YOUNG AND OLD

PAUL HALMOS

THE
DOLCIANI MATHEMATICAL EXPOSITIONS

Published by
THE MATHEMATICAL ASSOCIATION OF AMERICA

Dolciani Mathematical Expositions series design
ROBERT ISHI + DESIGN

TeX *macros*
RON WHITNEY

The Dolciani Mathematical Expositions

NUMBER TWELVE

PROBLEMS FOR MATHEMATICIANS, YOUNG AND OLD

PAUL HALMOS
Santa Clara University

Published and Distributed by
THE MATHEMATICAL ASSOCIATION OF AMERICA

Library of Congress Catalog Card Number 91-67386

Complete Set ISBN 0-88385-300-0
Vol. 12 ISBN 0-88385-320-5

Printed in the United States of America

Current printing (last digit):
10 9 8 7 6 5 4 3 2 1

The DOLCIANI MATHEMATICAL EXPOSITIONS series of the Mathematical Association of America was established through a generous gift to the Association from Mary P. Dolciani, Professor of Mathematics at Hunter College of the City University of New York. In making the gift, Professor Dolciani, herself an exceptionally talented and successful expositor of mathematics, had the purpose of furthering the ideal of excellence in mathematical exposition.

The Association, for its part, was delighted to accept the gracious gesture initiating the revolving fund for this series from one who has served the Association with distinction, both as a member of the Committee on Publications and as a member of the Board of Governors. It was with genuine pleasure that the Board chose to name the series in her honor.

The books in the series are selected for their lucid expository style and stimulating mathematical content. Typically, they contain an ample supply of exercises, many with accompanying solutions. They are intended to be sufficiently elementary for the undergraduate and even the mathematically inclined high-school student to understand and enjoy, but also to be interesting and sometimes challenging to the more advanced mathematician.

DOLCIANI MATHEMATICAL EXPOSITIONS

1. *Mathematical Gems,* Ross Honsberger
2. *Mathematical Gems II,* Ross Honsberger
3. *Mathematical Morsels,* Ross Honsberger
4. *Mathematical Plums,* Ross Honsberger (ed.)
5. *Great Moments in Mathematics (Before 1650),* Howard Eves
6. *Maxima and Minima without Calculus,* Ivan Niven
7. *Great Moments in Mathematics (After 1650),* Howard Eves
8. *Map Coloring, Polyhedra, and the Four-Color Problem,* David Barnette
9. *Mathematical Gems III,* Ross Honsberger
10. *More Mathematical Morsels,* Ross Honsberger
11. *Old and New Unsolved Problems in Plane Geometry and Number Theory,* Victor Klee and Stan Wagon
12. *Problems for Mathematicians, Young and Old,* Paul Halmos

PREFACE

I wrote this book for fun, and I hope you will read it the same way. It was fun indeed—the book almost wrote itself. It consists of some of the many problems that I started saving and treasuring a long time ago. Problems came up in conversations with friends, and in correspondence, and in books, and in lectures. I enjoyed them, thought about them, tried to solve them, tried to change them, and tried to think of new ones, and then I tried to organize and write down the ones I was fondest of—and this book is the result.

Problems, hints, solutions

The problems come complete with their statements, hints, and solutions.

The statements are not intended to be the most general ones known to humanity—their purpose is to stimulate thought, not abolish it. If you can think of extensions and improvements of the results asked for, you have done what I hoped you would do.

The hints are intended to be just that—suggestions intended to get you to look in a possibly profitable direction. A hint might, for instance, be a question. That doesn't mean that you have to know the answer to the question before you can solve the problem—but, possibly, the question might offer more food for thought than the original statement of the problem does.

As for the solutions, they are sometimes "wrong", or partially wrong —and then soon corrected. Don't give up if you see something wrong—keep going and try to set things right. The solutions make no pretense of being the best, the shortest, the most elegant, or even complete. Some of the solutions are chatty discussions, and some are hardly more than slightly extended hints that say just enough to point you in the right direction. In either case the purpose of a "solution" is to have you solve the problem—and to enjoy doing so.

Sources

The problem lover has many places to turn to—go to any library that has a mathematics section and your search for mathematical problem books is certain to be successful. That's one of the reasons why I do not offer a bibliography of problem sources, but I would like to mention a small number among the many famous and useful ones. Look for the names of Dörrie, Klambauer, Pólya and Szegő, and Steinhaus, and you will like everything you find. Look also at almost any issue of the *American Mathematical Monthly* or of the *Mathematics Magazine,* and, as long as we're talking about periodicals, don't forget *Scientific American,* especially the older issues in which Martin Gardner conducted his stimulating column.

Level

The level of this book, that is the mathematical sophistication of the audi ence it is intended to reach, is not sharply focused. The solutions of some of the problems require nothing more than that uncommon property known as common sense; those problems will make sense to (and can be solved by) high school students. Other problems require the maturity of a professional mathematician, who can be a second year graduate student or someone who has been earning a living by thinking about mathematics for more than twenty years.

Not only are the problems at many different levels in the book—their levels can vary even within one chapter. The order of difficulty is not a monotone function of where a problem occurs—a later problem in one chapter, or a later chapter, might well be easier than an earlier one.

Prerequisites

The subjects that the problems touch on can be roughly divided into eight parts, and here, by way of guidance to the prerequisites for the book, are the clue words and phrases that occur in them. Most of these words are not defined in the book, and especially some "well-known" words (example: "congruence") are not defined—if you don't know one of them, regard it as a challenge—guess, ask somebody, look it up—one way or another find out what it means.

Calculus: maxima and minima, the mean-value theorem, Newton's method, e and π.

Linear algebra: vector space, inner product, characteristic and minimal polynomial, eigenvector, Hermitian matrix, positive definite matrix, Jordan canonical form, skew-symmetric matrix.

Set theory: equivalence relation, countable and uncountable sets, power of the continuum, Hamel basis.

Topology: compact and connected, discrete topology, homeomorphism, product topology.

Analysis: limits, convergence, rational and irrational numbers, greatest lower bound, continuously differentiable, uniformly continuous, Cantor set and Cantor function, Picard's theorem, Borel set, Lebesgue measure, product measure.

Abstract algebra and number theory: groups, subgroups, cosets, permutations, fields and their characteristics, remainder theorem, isomorphism, inner automorphism, prime numbers, algebraic and transcendental numbers.

Probability: normal distribution, uniform distribution, random variable.

Functional analysis: Banach space, inverse mapping theorem.

The book contains a total of 165 problems, and no single concept is likely to occur in more than two or three of them—if there is one word in the list above that you don't know, you can probably still solve all the problems, and if there are twenty words that you don't know, you can still solve well over half of them.

Sometimes a word is just casually mentioned, in order to point out a possible connection of one subject with another. (Possible example: "The theorem can be generalized to Banach spaces.") This is particularly likely to happen in the comments appended to some of the problems and solutions, and it might be of interest to you if you are an expert in one of the subjects but not in the other. If, however, you are not an expert in either subject, and you run across a casual mention in which you don't know the crucial word, don't let that intimidate you—just skip it.

Notation

The notation and the terminology I use are essentially standard—when a choice had to be made I made it, and, of course, you are at liberty to change mine and make your own. There are, however, three items that you should be explicitly warned about.

1. I am prejudiced about log—as far as I am concerned it stands for the so-called natural logarithm, logarithm to the base e, and nothing else. An alternative that played a role for a while, logarithm to the base 10, is an unnatural and by now unnecessary abomination.

2. In linear algebra scalar matrices occur frequently. The most common one is 0, almost always denoted, with no danger of confusion, by the same symbol as is used for the number. The next most common one is the identity matrix (corresponding to the identity transformation). Notational consistency demands that that matrix (transformation) be denoted by the symbol 1 (the same as for the number), and I do so. Correspondingly, if λ is a scalar, the corresponding scalar matrix, the multiple of the identity by the scalar λ, is denoted by the symbol λ—the same as for the number.

3. If X and Y are sets, then the relative complement of Y in X is denoted by $X - Y$, the traditional symbol for subtraction. Some purists object and point to the conflict of that notation with one that occurs in additive groups: if X and Y are subsets of such a group, then the set of all elements of the form $x - y$, with x in X and y in Y, should be denoted by $X - Y$. The purists are right, and on

one or two occasions I yield to their demands, reluctantly and with appropriate warnings—the rest of the time set subtraction means relative complement. When both relative complement and group subtraction occur in the same paragraph, they are denoted by $X \setminus Y$ and $X - Y$ respectively.

Credits

As for credits—who discovered what?, who was first?, whose solution is the best?—I do not give any, not even when I am sure I know. I keep silent even in the one or two cases in which I was present at the discovery of a problem or a solution—and, all the more, I keep silent when no one really knows, when a customary assignment of credit is difficult or impossible to verify. It wouldn't be fair to give credit in some cases and not in others. We are all pretty sure that Pythagoras did not discover the Pythagorean theorem, and it doesn't really matter, does it? The beauty of the mathematics speaks for itself, and so be it.

Thanks

Books are rarely written by just one person, and this one certainly was not. Friends (Sheldon Axler, Woody Dudley, and Pete Rosenthal) gave me advice and help and answers to my bewildered questions, and my friend and boss, Jerry Alexanderson, gave me time off from other duties to enable me to finish the book in a finite time. I am grateful to them.

Degrees of gratitude are hard to measure, but I think I can say that I am especially grateful to three others, the "official" readers of the manuscript, without whom this would be a different book, one that I couldn't like as much as I want to. Kati Bencsáth, John Ewing, and Abe Hillman read every word of one or another version of the work and made many suggestions for improvements. They found errors, they found notational collisions, they found unclear explanations, and in each case they told me what I should do. I accepted many of their suggestions and refused to accept others, and, worse yet, I made changes after their job was finished. It is quite possible that if I had always done what they said, this would be a better book, but, especially in stylistic matters, I found myself stubborn, and, consequently, the faults that you are likely to encounter are mine alone. Nevertheless, I am grateful to them, truly and greatly grateful—thanks a lot.

Contents

CHAPTER 1

COMBINATORICS

1 A. Tennis matches

1 A

Sometimes the main challenge of a problem is not just to find a solution, but to find the "right" solution, the beautiful one, the one that gets to the heart of the matter. Here is an example of a problem that everybody can solve, one way or another. If your life depended on solving it, you would be in no danger so long as you had adequate patience and were willing to add up a lot of numbers. The real problem, however, is to find the reason why the final number is the right one.

> **Problem 1 A.** *Suppose that* 1025 *tennis players want to play an elimination tournament. That means: they pair up, at random, for each round; if the number of players before the round begins is odd, one of them, chosen at random, sits out that round. The winners of each round, and the odd one who sat it out (if there was an odd one), play in the next round, till finally there is only one winner, the champion. What is the total number of matches to be played altogether, in all the rounds of the tournament?*

Comment. The problem is classified under "combinatorics", but no one really knows what that word means. It vaguely indicates something that has no precise definition and probably needs none—in this respect it is like other words such as "algebra" or "loyalty". Usually there is universal agreement that something can be classified under combinatorics, but disagreements are possible (just as with loyalty). Most often combinatorics has to do with finite sets and the questions it asks are about counting ("how many?"). Such questions can be easy and they can be fiendishly difficult; their answers can be dull and they can be breathtakingly beautiful.

1 B

1 B. Friends and strangers

Suppose that the use of words is perverted slightly by saying that any two human beings are either "friends" or "strangers". It is to be understood that friendship is a symmetric relation: if X is a friend of Y, then Y is a friend of X. In this language, for a randomly chosen bunch of three people it is quite likely that a couple of them are friends and another couple are strangers; it is possible, though not likely, that they are all friends or else all strangers. For any four people, similarly, it is easily possible that they contain a pair of friends and also a pair of strangers; it is also possible, though less likely, that there are three among them who are all friends or else all strangers.

For five people things get a little harder to count. Is it possible to have acquaintance relations among five people so that no three of them are all friends and at the same time no three of them are all strangers? That takes a little thought. The answer turns out to be yes. Indeed, think of five people seated around a circular table, and suppose that any two in neighboring seats are friends, but any two not in neighboring seats are strangers. In that case, no matter how three of them are selected, it must be the case that some two of them are neighbors and some two

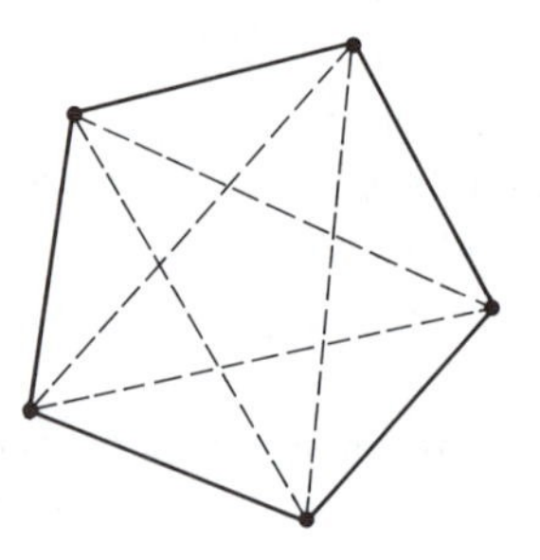

FIGURE 1

are not; in other words it is never the case that three are all friends and never the case that three are all strangers.

The relations here discussed can also be described geometrically. Consider some finite set of points in space (three-dimensional Euclidean space) with line segments joining each pair, and suppose that each such segment is painted either red or blue. If points are thought of as representing people, and the red and blue segments joining pairs of them are thought of as representing friends and strangers respectively, then the geometric version of the friends-and-strangers question becomes this: do there exist three segments that form a triangle whose sides are all of the same color? A properly chosen technical word makes for a shorter and clearer formulation: does there exist a *monochromatic* triangle? If the number of points is three, or four, or five, then the preceding paragraphs show that the answer may be no: monochromatic triangles may fail to exist. What happens as the number of points increases—does the existence of monochromatic triangles become more or less likely?

> **Problem 1 B.** *Is there a positive integer n such that if each of n points is joined to every other one by either a red or a blue segment, then, no matter how the colors of the segments are chosen, a monochromatic triangle always exists?*

Comment. A *no* answer demands a proof, of course, and a *yes* answer an example, but for a problem like this one a *yes* answer would demand more than just pointing at a number. Which numbers could do the trick?—how many of them could there be?—what could the smallest one be?

1 C. Symmetric matrices

1 C

Sometimes a matrix is defined to be "a square array", but that misses much of the point, just as defining a library to be many sheets of paper misses the point. For this problem, however, a matrix is nothing but a square with numbers in it; none of the algebraic theory of matrices is needed to solve it.

A matrix whose entry in row i and column j is denoted by a_{ij} is called *symmetric* in case $a_{ij} = a_{ji}$ for each i and j. If each row of a symmetric $n \times n$ matrix is a permutation of the numbers $\{1, 2, \ldots, n\}$, then the same thing is true of each column. That's obvious, isn't it? Flipping a matrix over the main diagonal always interchanges rows and columns, and symmetry implies that such a flip doesn't change anything. (The *main diagonal* of a matrix (a_{ij}) is the sequence of numbers

$$\{a_{11}, a_{22}, \ldots, a_{nn}\},$$

and to "flip" means to replace the matrix by the new matrix (b_{ij}), where $b_{ij} = a_{ji}$ for each i and j.) Question: if each row of a symmetric $n \times n$ matrix is a permutation of the numbers $\{1, 2, \ldots, n\}$, is the same thing true of the diagonal? The answer is no; a trivial example is the matrix $\binom{1\,2}{2\,1}$. A less trivial example is

$$\begin{pmatrix} 1 & 2 & 3 & 4 \\ 2 & 1 & 4 & 3 \\ 3 & 4 & 1 & 2 \\ 4 & 3 & 2 & 1 \end{pmatrix}$$

Problem 1 C. *For which positive integers n can there exist a symmetric $n \times n$ matrix such that each row is a permutation of $1, \ldots, n$ (and therefore so is each column), but the diagonal is not?*

1 D

1 D. The marriage problem

Problem 1 D. *Suppose that each of a set of 1000 boys is acquainted with some (possibly none) of a set of 7000 girls. Under what conditions on the acquaintanceships is it possible for each boy to marry one of his acquaintances?*

Comment. The difficulty of the problem is not to solve it (that's not so hard), but to interpret it. What does it mean to ask for "conditions on the acquaintanceships"?

Obviously some conditions must be satisfied: for instance there must be no hermit among the boys. That is, the permissive phrase "possibly none" had better not apply: a boy who knows no girls cannot get married. In other words, a necessary condition is that each boy be acquainted with at least one girl. There is another condition that must clearly be satisfied collectively by the set of all boys; among them all they had better have at least 1000 girl friends. If they don't, if, for instance, each boy knows 900 girls, but it's always the same 900 for all the boys (so that 6100 of the girls know no boys at all), then, once again, no legal mass marriage is possible.

Other things can go wrong, just like the two possibilities already mentioned but numerically more stringent. If, for instance, there are two shy boys, each of whom has managed to meet only one girl, the same girl for both, the marriage distribution is ruined. Consequence: each set of two boys must have at least two girl friends between them. Similarly, each set of three boys must, among them, have at least three girl friends, etc., etc.; each set of 500 boys must, among them, have at least 500 girl friends, etc., etc. In one sentence: for each k, between 1 and 1000 inclusive, it is necessary that each set of k boys collectively be acquainted with at least k girls.

That's a lot of different conditions that the acquaintanceship pattern of the boys is required to satisfy before the marriage problem can have a solution, but possibly, even so, they are not sufficient to guarantee a solution. For all we know, all these conditions can be satisfied and still a marriage arrangement cannot be worked out for other, subtler, reasons. The only way to learn the truth of the matter is either to exhibit an acquaintanceship pattern that satisfies the stated conditions but still admits no solution to the marriage problem, or else to prove that when the stated conditions are satisfied, then the marriages can always be arranged.

1 E. Marital infidelity

1 E

Having solved the marriage problem, we are ready to turn to a study of what happens after that.

> **Problem 1 E.** *Every man in a village knows instantly when another's wife is unfaithful but never whether his own is. Each man is completely intelligent and knows that every other man is. The law of the village demands that when a man can PROVE that his wife has been unfaithful, he must shoot her before sundown the same day. Every man is completely law-abiding. One day the mayor announces that there is at least one unfaithful wife in the village. The mayor always tells the truth, and every man believes him. If in fact there are exactly forty unfaithful wives in the village*

(but that fact is not known to the men), what will happen after the mayor's announcement?

Comment. Since every man knew all along that there are unfaithful wives, what difference can the mayor's announcement possibly make?

Note. Down the road from this mythical village is another one in which the roles of men and women are interchanged.

1 F

1 F. Monotone subsequences

A finite sequence $\{a_1, a_2, \ldots, a_n\}$ of real numbers is called monotone increasing if $a_1 < a_2 < \cdots < a_n$, monotone decreasing if $a_1 > a_2 > \cdots > a_n$, and just simply monotone if it is either monotone increasing or monotone decreasing. For sequences of length one the definition doesn't make much sense, and sequences of length two are always monotone, provided only that the two terms a_1 and a_2 are distinct. Monotone sequences always have distinct terms, and therefore a discussion that tries to find or to avoid monotone sequences might as well restrict attention to sequences with distinct terms only, and that's what will be done from now on.

Easy examples such as $\{1, 3, 2\}$ show that sequences of length three or more may fail to be monotone, but each of their subsequences of length two is monotone, just because it has length two. (Emphasis: a "subsequence" is a sequence formed from a given one by throwing away some terms, but maintaining the order of the others.) Put it another way: monotone subsequences of length two are hard to avoid. Question: can longer monotone subsequences be avoided? For example: do sequences of length four necessarily have monotone subsequences of length three? It doesn't take much experimentation to reveal that the answer is no;

$$\{2, 1, 4, 3\}$$

is a sequence of length four that has no monotone subsequences of length three. Why is that? Is it because four is too short—there is not enough room? Could it be that sequences of length five always do have monotone subsequences of length three?

The answer to the last question demands a little more experimentation, and the result may be a surprise: it turns out that the answer is yes. Once that is granted, then it follows trivially that every sequence of length greater than or equal to five must have monotone subsequences of length three. Is there a general phenomenon hiding here? Sequences of length five do not have to have monotone subsequences of length four; one example is

$$\{3, 2, 1, 5, 4\}.$$

What about sequences of length six, or seven, or eight, etc.?—do they have to have monotone subsequences of length four? Could it be that there are sequences of arbitrarily great length without monotone subsequences of length four? And what are the facts for lengths greater than four—are they the same, or do they change as greater lengths are allowed to enter the competition?

Problem 1 F. *True or false: to every positive integer n there corresponds a length such that every sequence of that length or more (with all distinct terms) must contain a monotone subsequence of length n?*

1 G. Increasing rows

1 G

Suppose that a finite set of distinct numbers is arranged in a rectangular array in such a way that each row forms an increasing sequence. Possible example:

$$\begin{array}{ccc} 21 & 25 & 28 \\ 17 & 18 & 27 \\ 20 & 24 & 26 \\ 19 & 22 & 23 \end{array}.$$

Then, one after another, rearrange each column so that the the new columns form increasing sequences (as the columns are traversed from top to bottom). The result in the case of the example is

$$\begin{array}{ccc} 17 & 18 & 23 \\ 19 & 22 & 26 \\ 20 & 24 & 27 \\ 21 & 25 & 28 \end{array}.$$

A curious thing has happened: even though the rows were ignored in the course of the rearrangement, it turned out, in the final result, that they ended up as increasing sequences again. Is that an accident, is that a special property of the numbers at hand (which happen to be consecutive positive integers), or is it always true?

Problem 1 G. *If each row of a rectangular array of distinct real numbers is an increasing sequence, is the same thing true of the rectangular array obtained by rearranging each column to be an increasing sequence (read from top to bottom)?*

1 H

1 H. Handshakes

Problem 1 H. *My wife and I were invited to a party attended by four other couples, making a total of ten people. As people arrived, a certain amount of handshaking took place in an unpredictable way, subject only to two obvious conditions: no one shook his or her own hand and no husband shook his wife's hand. When it was all over, I became curious and I went around the party asking each person: "How many hands did you shake? ... And you? ... And you?" I asked nine people (everybody, including my own wife), and I was interested to note, and I hereby report, that I got nine different answers.*

How many hands did my wife shake?

1 I

1 I. Seating

Suppose that four people, numbered 1, 2, 3, 4, are seated in four chairs, numbered 1, 2, 3, 4, like this:

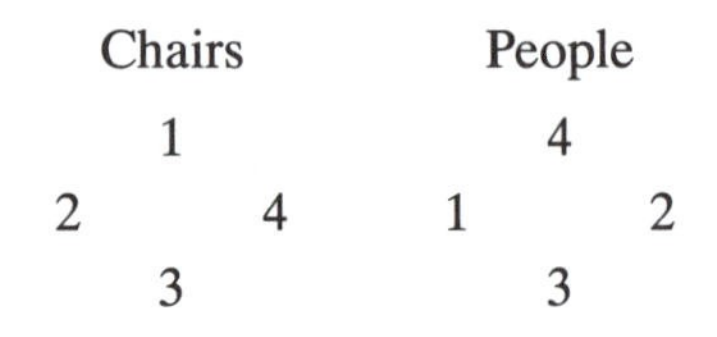

Note that person number 3 is sitting in chair number 3. Another way of seating the same people, in the same circular order, on the same chairs (which are left fixed) is this:

Chairs			People		
	1			2	
2		4	4		3
	3			1	

In this case no person bears the same number as the chair he is in. Can that always be done?

Problem 1 I. *Can a thousand people seated around a circle in seats numbered from* 1 *to* 1000*, each person bearing one of the numbers from* 1 *to* 1000*, be re-seated so as to preserve their (circular) order and so that no person's number is the same as that of his chair?*

1 J. Unending chess

1 J

In an attempt to guarantee that no game of chess will keep going forever, the following rule was once proposed: the game is a draw if some finite sequence of moves is repeated three times in a row. Does the rule work?

Without some such rule, it is clear that a game of chess could continue ad infinitum. There exist chess problems of this kind: given a certain position, if it's White's turn, then he can win in two moves, but if it's Black's turn, then he can make a sequence of forcing moves that after a while produce the same position as the one where he started, and by doing that over and over again he can produce an infinite game. More trivially: if two stubborn players want, for some reason, to monopolize a chessboard, they can do so ad infinitum. Example: White starts by moving a knight, Black replies by a similar move, White returns the moved knight to its starting position, Black does the same, and they keep on alternating this way.

The obvious interpretation of the question is this: is it true that if the proposed rule is added to the other (standard) rules of chess, then neither normal play nor stubborn conspiracy can produce an infinite game? In other words, is it true that, with the proposed rule, each legal game of chess would have to end (win, lose, or draw) in a finite number of moves?

It would be an extraordinarily boring task to devise a notation in which each possible position of the pieces on a chessboard is indicated by a symbol, but it could be done. Once that's done, a game of chess could be described by a sequence of symbols; the rules of the game would describe which sequences can occur. The proposed rule states that if in a legal sequence some finite block of some length occurs three times in succession, then at the end of the third repetition of the block the sequence should be ended and the game should be declared a draw.

To study the effectiveness of the proposed rule, it is not really necessary to consider sequences whose terms are drawn from a large set of complicated symbols. Using only a small amount of mathematical ingenuity, a code could be devised that replaces each symbol by a finite sequence of 0's and 1's. The question becomes therefore a question about (possibly infinite) sequences of 0's and 1's.

Problem 1 J. *Does there exist an infinite sequence of* 0*'s and* 1*'s in which no finite block of any length occurs three times in succession?*

1 K

1 K. Surgical gloves

When a surgeon operates, he must protect his patient from infections, and, at the same time, protect himself—those two jobs are what sterile gloves are designed to accomplish.

Problem 1 K. *Can a surgeon with only two pairs of sterile gloves perform operations on three distinct patients and keep everybody safe?*

CHAPTER 2

CALCULUS

2 A. Cosine iterated

2 A

Functions can sometimes be composed and, in particular, can sometimes be iterated. To compose two functions f and g means to form the function h defined by

$$h(x) = f(g(x)),$$

or, in other words, to form the "function of a function" that they determine. That expression is sloppy—it does not take into account various important delicacies.

In the first place, in order that the function h be formable at all it is clearly necessary that the domain of definition of the function f include the values that the function g takes on—the substitution into f of a value such as $g(x)$ must make sense. More elegantly said, in order that h be formable it is necessary that

$$\text{ran } g \subset \text{dom } f.$$

Another important observation is that the composition depends on the order in which it is performed; even if $f(g(x))$ makes sense, $g(f(x))$ may fail to make sense. The function h defined above is denoted by the symbol

$$f \circ g,$$

the intention being that the small circle be reminiscent of a dot indicating multiplication. Note that to apply $f \circ g$, the composition of f and g, the function g must be applied first and then f—it works, and is to be read, from right to left. That's not an artificial condition imposed from outside—it is in the nature of things, implied by the way functions are customarily thought of and written. The application of g to x is indicated by writing g to the left of x, as in $g(x)$, and, consequently, the symbol $(f \circ g)(x)$ indicates that the factor g next on the left to x is what's applied to x, and to which, in turn, f is then applied.

If a function f is such that its range is included in its domain, if, in other words, it transforms its domain into a part of that domain, then the composition $f \circ f$ makes sense; it is usually denoted by f^2. The exponent indicates "multiplication" in the sense of function composition, not numerical multiplication. If confusion is likely to arise, a modified symbol such as $f^{(2)}$ can be used.

What can be done once, can be repeated; if the basic condition (range included in domain) is satisfied, then the function f can be composed with itself over and over again, yielding f^3, f^4, etc. The procedure of "iteration" so described arises frequently in mathematics and is frequently useful; it produces some surprising phenomena even with familiar elementary functions such as cos (cosine) on its natural domain $[-\pi, +\pi]$. Note that cos transforms the closed interval $[-\pi, +\pi]$ into the closed interval $[-1, +1]$, so that its range is included in its domain—iteration makes sense.

Problem 2 A. *What happens when cos is iterated infinitely often —in other words what properties can the sequences of numbers such as $\{\cos^{(n)} x\}$ have?*

Comment. This is the first problem in the calculus section—why is it there? What is calculus about anyway? The answer is that calculus studies certain properties of certain functions, and, in particular, it studies limits of sequences and sums of series, it studies derivatives and integrals, and it studies maxima and minima. Those are the concepts that the problems in this section will consider, very briefly to be sure.

The function cosine first arises for most of us in elementary trigonometry in expressions such as $\cos x$, where x is a certain number of degrees that measure an angle (whatever that may be). After a while we learn to measure angles in radians and get used to the fact that $\cos x$ makes sense for any real number x; in particular the value of $\cos x$, which is surely not an angle, can be substituted into cos.

2 B. Square root limits

2 B

What does the sequence of numbers

$$\sqrt{n(n+1)} - n$$

do as $n \to \infty$? Does it tend to a finite limit or does it diverge to ∞? What about the sequence of numbers

$$\sqrt{2n(n+1)} - n?$$

The answers are probably not obvious at first glance, and the result might be a bit of a surprise. It turns out that the first sequence converges to the limit $\frac{1}{2}$ whereas the second becomes infinite.

The same technique can be used to prove both answers: multiply the given expression by 1, written in the first case as

$$\frac{\sqrt{n(n+1)} + n}{\sqrt{n(n+1)} + n}$$

and in the second case as

$$\frac{\sqrt{2n(n+1)} + n}{\sqrt{2n(n+1)} + n.}$$

Indeed, in the first case the product simplifies to

$$\frac{n^2 + n - n^2}{\sqrt{n^2+n} + n} = \frac{1}{\sqrt{1 + \frac{1}{n}} + 1},$$

and in the second case to

$$\frac{2n^2+2n-n^2}{\sqrt{2n^2+2n}+n} = \frac{n+2}{\sqrt{2+\frac{2}{n}}+1}.$$

What general fact is lurking behind these two examples?

Problem 2 B. *Which of the numbers* 1, 2, $\sqrt{3}$, e^4, π^5 *can be the limit of a sequence of numbers of the form* $\sqrt{n}-\sqrt{m}$?

2 C

2 C. Irrational multiples

Since $\sqrt{2}$ is irrational, a number of the form $p+q\sqrt{2}$, where p and q are integers and $q \neq 0$, can never be 0—but can it be nearly 0? Or is there a lower limit such that no number of that form can be smaller than that?

Problem 2 C. *Which real numbers are limits of numbers of the form* $p+q\sqrt{2}$, *where* p *and* q *are integers?*

2 D

2 D. Square roots as limits

Speaking of square roots, how can they be calculated? Grade school children used to be taught an algorithm for finding square roots, but that went out of fashion even before the ubiquity of hand held calculators. (Are long multiplication and long division going the same way?) If you were stuck on a desert island and couldn't get off till you found the square root of 303, what would you do?

If t is a positive number and s is its positive square root, then $s = \frac{t}{s}$—that's not a profound statement. Suppose, however, that t is known but s is not, and that an emergency arises in which it is necessary at least to guess s. If the guess is too small, then $\frac{t}{s}$ is too large, and vice versa; in either case it seems reasonable to believe that the average of s and $\frac{t}{s}$ is a better guess than either s or $\frac{t}{s}$. That observation is often correct, and suggests an algorithm for finding square roots. Such inspired suggestions sometimes work; is this one of those times?

Problem 2 D. *If* s *and* t *are positive numbers, and if a sequence* $\{x_n\}$ *is defined by* $x_0 = s$ *and*

$$x_{n+1} = \frac{1}{2}\left(x_n + \frac{t}{x_n}\right),$$

for $n = 0, 1, 2, \ldots,$ *for which values of s and t does the sequence converge to a limit, and, when it does, what is the limit?*

Comment. For each t there is at least one value of s for which convergence does take place, namely $s = \sqrt{t}$; in that case $x_n = s$ for every value of n.

2 E. A maximum

2 E

The determination of maxima and minima is one of the most important applications of calculus—and it is important to be aware that problems of that sort can arise even when calculus does not enter. Some thought before applying the cast-iron rule "set the derivative equal to zero" is frequently in order.

Problem 2 E. *What is the maximum value of*

$$|ax + ay + bx - by|$$

over the "unit cube" in 4-dimensional space determined by

$$|a| \leqq 1, \quad |b| \leqq 1, \quad |x| \leqq 1, \quad |y| \leqq 1?$$

2 F. Double integral

2 F

Sometimes definite integrals need to be evaluated even when the corresponding indefinite ones cause extreme difficulties. That can happen with the so-called "elementary" functions, and it can even happen that their indefinite integral is not just difficult but, in fact, impossible to evaluate. What that means is that the anti-derivative (the indefinite integral) is not an elementary function at all. The most famous example is probably the one that occurs in the normal distribution in probability theory; it is the function f defined by

$$f(x) = e^{-x^2}.$$

An expression such as $\int_{-\infty}^{x} e^{-t^2}\,dt$ makes perfectly good sense, and it defines a function g whose derivative is exactly the function f, but g can be proved to be *not* an elementary function, and there is nothing that calculus can do about that. Nevertheless, the definite integral $\int_{-\infty}^{+\infty} e^{-x^2}\,dx$ can be evaluated—there is a simple trick. If

$$\int_{-\infty}^{+\infty} e^{-x^2}\,dx = p,$$

then

$$\begin{aligned} p^2 &= \left(\int_{-\infty}^{+\infty} e^{-x^2}\,dx\right)\cdot\left(\int_{-\infty}^{+\infty} e^{-y^2}\,dy\right) \\ &= \int_{-\infty}^{+\infty}\int_{-\infty}^{+\infty} e^{-(x^2+y^2)}\,dx\,dy, \end{aligned}$$

which, written in polar coordinates, says that

$$p^2 = \int_0^{2\pi}\int_0^{\infty} e^{-r^2} r\,dr\,d\theta = \int_0^{2\pi} d\theta \int_0^{\infty} e^{-r^2} r\,dr.$$

Since the evaluation of the integral with respect to r is an easy calculus exercise—its value is $\frac{1}{2}$—it follows that $p^2 = \frac{1}{2}\cdot 2\pi = \pi$, and hence that

$$\int_{-\infty}^{+\infty} e^{-x^2}\,dx = \sqrt{\pi}.$$

The following problem is similar in that it firmly resists attacks via indefinite integration.

Problem 2 F. *Is the value of the integral*

$$\int_0^1\int_0^1 \frac{1}{1-xy}\,dx\,dy$$

more than 1 *or less? What about* 2? 1.5? 1.25? 1.75?

2 G

2 G. Product rule

Students sometimes complain that the rules they learn in calculus are too complicated. Why do we make $(fg)'$, the derivative of a product, so complicated?—why don't we just let it be the obvious thing, namely $f'g'$? In case both f and g are constant (not necessarily the same constant), then both $(fg)'$ and $f'g'$ are 0 and all is well. Is that the only time that happens? The only thing that's obvious is that if it ever happens, that is if f and g are functions such that $(fg)' = f'g'$, then the replacement of both f and g by constant multiples of themselves produces another pair of fuctions satisfying the same equation.

Problem 2 G. *Are there many pairs of functions f and g such that $(fg)'$ and $f'g'$?*

2 H. Exponential inequality

2 H

Is it obvious that $e^x > 1+x$ for all real values of x? It had better not be: if $x = 0$, then the inequality in question becomes the equation $1 = 1$. What about the x's different from 0? In that case the inequality is true, and whether it's regarded as obvious depends on how one thinks of the exponential function.

Think of the picture: the graph of $y = e^x$ goes through the point $(0, 1)$, ascends sharply as x increases, and descends (but never below the x-axis) as x decreases. Since the slope at $x = 0$ is 1 (the value of the derivative at 0), the equation of the tangent line at $(0, 1)$ is $y = 1 + x$, and the picture shows the inequality clearly. That's not really a proof of anything, but it is at least a convincing reminder.

If e^x is interpreted as the infinite series

$$1 + x + \frac{x^2}{2!} + \frac{x^3}{3!} + \cdots,$$

then it is indeed obvious that $e^x > 1 + x$ for all positive values of x, but the negative ones require another look at the very least.

One way to look is to use standard calculus: minimize $e^x - x - 1$. The first derivative is $e^x - 1$, which vanishes at $x = 0$ only (only real x's are under consideration); the second derivative is e^x, which is positive, so that the value at 0, which is 0, is a minimum. Conclusion: $e^x - x - 1 > 0$ for all real x, except 0, as desired.

Often the best way to think of e^x is as the limit of $\left(1+\frac{x}{n}\right)^n$ (as $n \to \infty$), but that doesn't seem to make the inequality any more obvious for either positive or negative values of x. It is, nevertheless a good way to approach an elementary and correct proof of the inequality; here is how that goes.

Begin with the auxiliary inequality that says

$$p^n > 1 + n(p-1)$$

for every positive integer n and every positive number p (except $p = 1$, in which case it degenerates to an equality). To prove that inequality, recall that

$$\frac{1-p^n}{1-p} = \frac{p^n-1}{p-1} = 1 + p + p^2 + \cdots + p^{n-1}$$

whenever $p \neq 1$. It follows that

$$\frac{p^n-1}{p-1} > n$$

whenever $p > 1$ and

$$\frac{1-p^n}{1-p} < n$$

whenever $p < 1$. In each of these two cases the asserted auxiliary inequality is an immediate consequence.

Suppose now that x is an arbitrary non-zero real number. If n is sufficiently large, then $\left|\frac{x}{n}\right| < 1$, and, consequently, $1 + \frac{x}{n} > 0$. Apply the auxiliary inequality to $1 + \frac{x}{n}$ in the place of p, and infer that

$$\left(1+\frac{x}{n}\right)^n > 1 + x.$$

Go to the limit as $n \to \infty$ and conclude that $e^x \geqq 1 + x$, as desired.

That wasn't hard, but it was a bit of trouble—possibly more trouble than one could have expected. How much more trouble is caused by using a base other than e?

Problem 2 H. *For which positive real numbers a is it true that*

$$a^x \geqq 1 + x$$

for all real values of x?

2 I. Railroad track

2 I

Problem 2 I. *A railroad track runs absolutely straight and level for exactly one mile (the curvature of the earth having been flattened). Assume that its two ends remain fixed, but one additional foot of track is inserted in the middle and then seamlessly welded in with the rest. Assume, moreover, that the track buckles up in the shape of an arc of a circle. Question: how far is the middle of that arc off the ground?*

Comment. Figure 2, grossly the wrong scale, illustrates the situation.

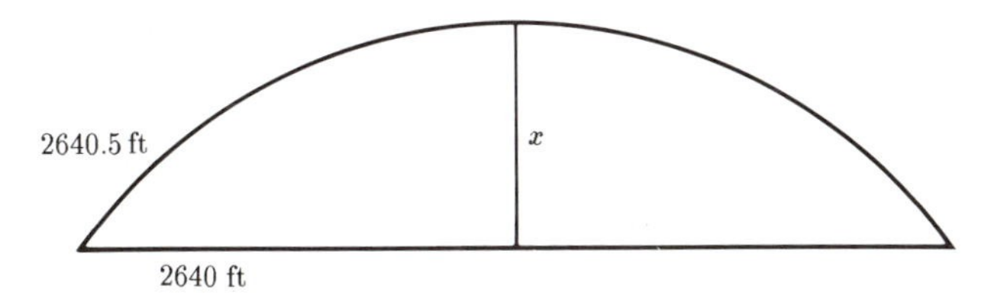

FIGURE 2

Question: what is the value of x? What is a reasonable guess? Is x very small (like $\frac{1}{100}$ of an inch), or is it very large (like 50 feet), or is it somewhere between (like a foot or two)?

2 J. Acceleration

2 J

Problem 2 J. *If it is known that a particle moves always from left to right on a straight line, starts and ends at rest, and covers a* *unit distance in a unit time, what can be said about the magnitude of its acceleration?*

Comment. In the language of pure mathematics the question should be phrased in some such manner as this: if a real-valued function x on the unit interval is twice continuously differentiable, monotone, and such that $x(0) = 0$, $x(1) = 1$, $x'(0) = x'(1) = 0$, then what can be said about $|x''(t)|$? Is the "translation" clear? The idea is that for each t in $[0, 1]$ the value $x(t)$ is the position of the particle at time t. The condition of being "twice differentiable" is demanded by the very fact that the principal subject is acceleration. "Twice continuously differentiable" is a mathematical convenience—possibly something could be said even if the second derivative is not continuous, but the central idea is likely to be clearest in the continuous case. "Monotone" just says that the particle "moves always from left to right", the restrictions on $x(0)$ and $x(1)$ translate "covers a unit distance in unit time", and the restrictions on $x'(0)$ and $x'(1)$ translate "starts and ends at rest". Note that the question concerns the absolute value of x''—that is "the magnitude" of the acceleration.

Since the acceleration is a continuous function, it is bounded; what can its least upper bound be? Could the motion be arranged so that the magnitude of the acceleration is never greater than 1? That's the sort of thing that the question is about.

CHAPTER 3

PUZZLES

Introduction

The problems in this section are less technical, more elementary, (and therefore more fun?) than most. They are not really problems, but puzzles; the kind of puzzles that you might see in the Sunday supplement of your local paper, and that the cop on the beat, your grocery bagger, and even your lawyer can enjoy and solve. Here, to begin with, is a silly little problem that has a certain amount of charm anyway. It is easy—anybody can solve it—but the right way to look at it is likely to come to most people on the second try, not the first, and the answer is one that most people find surprising.

3 A. Cucumbers

Problem 3 A. *Cucumbers are assumed, for present purposes, to be a substance that is 99% water by weight. If 500 pounds of cucumbers are allowed to stand overnight, and if the partially evaporated substance that remains in the morning is 98% water, how much is the morning weight?*

3 B

3 B. Irrational exponents

The number 2 is rational, and so is the number $\frac{1}{2}$ but $2^{1/2}$ is not. That is: there exist rational numbers α and β such that α^β is irrational. Can it be done the other way around?

> **Problem 3 B.** *Do there exist irrational numbers α and β such that α^β is rational?*

3 C

3 C. Million factorial.

> **Problem 3 C.** *How many zeroes does* 1,000,000! (*one million factorial*) *end in*?

Comment. It should be and almost certainly is needless to say that the question refers to the ordinary decimal notation for integers.

3 D

3 D. Coconuts

> **Problem 3 D.** *After gathering a pile of coconuts one day, five sailors on a desert island agree to divide them evenly after a night's rest. During the night one sailor gets up, divides the nuts into five equal piles with a remainder of one, which he tosses to a conveniently nearby monkey, and, secreting his pile, mixes up the others and retires. The second sailor does the same thing, and so do the third, the fourth, and the fifth. In the morning the remaining pile of coconuts* (*less one*) *is again divisible by* 5. *What is the smallest number of coconuts that the original pile could have contained*?

3 E

3 E. Three 2's

A once popular kind of arithmetic puzzle was to start with a "few" numbers and construct "many" by forming arithmetic combinations. Example: construct as long an initial segment as possible of the set of natural numbers using exactly four 3's for each term. Possible beginning:

$$\frac{3+3}{3+3}, \quad \frac{3}{3}+\frac{3}{3}, \quad \frac{3+3+3}{3}.$$

The problem is not perfectly clear. What operations are allowed? Are, for instance, powers allowed, as in $3+(\frac{3}{3})^3$? Are roots allowed, as in $3+\sqrt[3]{3^3}$? Is (decimal) juxtaposition allowed, as in $\frac{33}{3}-3$?

Problem 3 E. *Which positive integers can be expressed using only three 2's and "elementary operations"?*

3 F. Magic squares

3 F

What is a magic square? According to the simplest possible definition it is a square array of numbers, or, in standard mathematical terminology, a matrix, with the property that the the row sums and the column sums are all the same. If the size of the matrix is $n \times n$, then that comes to $2n$ sums that are all supposed to have the same value. Sometimes it is demanded that the two main diagonals, forward and backward, also have that same sum, but that is a special case; in what follows that extra demand will not be made.

What kinds of numbers are allowed to enter a magic square? The answer is a matter of taste. In most of the popular examples the entries are restricted to positive integers only, but matters become a little more mathematically amenable if, as in the discussion below, all integers are allowed. (You can replace the additive group of integers by an arbitrary abelian group if you like.) So, for example,

$$\begin{pmatrix} 1 & 6 & 8 \\ 9 & 2 & 4 \\ 5 & 7 & 3 \end{pmatrix}$$

is a magic square, and so also are

$$\begin{pmatrix} 1 & 2 & 5 \\ 3 & 4 & 1 \\ 4 & 2 & 2 \end{pmatrix} \quad \text{and} \quad \begin{pmatrix} 1 & 6 & 3 \\ 9 & 2 & -1 \\ 0 & 2 & 8 \end{pmatrix}.$$

The first of these latter examples calls attention to the fact that repeated entries can enter, and the second to the possibility of zero and negative entries. Is there any structure hiding here? How can these, and other examples, be generated? Trying to construct a magic square, how long can we put in numbers at random before the freedom disappears and our hand is forced?

Problem 3 F. *Which* $(n-1)\times(n-1)$ *matrices can appear as top left corners of* $n \times n$ *magic squares?*

3 G

3 G. Consecutive integers

The equations

$$5 = 2 + 3,$$

$$10 = 1 + 2 + 3 + 4,$$

$$12 = 3 + 4 + 5$$

show that each of the numbers 5 and 10 and 12 can be written as a sum of two or more consecutive positive integers. Is 18 equal to such a sum? What about 16?

Problem 3 G. *Which positive integers are sums of two or more consecutive positive integers?*

CHAPTER 4

NUMBERS

4 A. Powers and primes

4 A

If $n = 1$, then $n^4 + 4^n = 5$, a prime. Is that a common phenomenon or a rare accident?

Problem 4 A. *For which positive integers n is $n^4 + 4^n$ a prime?*

4 B. Primes of 1's and 0's

4 B

We can all tell, just by looking, that 10, and 100, and 1,000 are not primes, and that 7895 is divisible by 5. Usually, however, the decimal notation for a positive integer does not give much information about divisibility properties.

Problem 4 B. *Which of the positive integers*

$$101, \qquad 10101, \qquad 1010101, \ldots,$$

with alternating 0's and 1's beginning and ending with 1, can be primes?

4 C

4 C. Chinese remainder theorem

We know how to solve an equation such as $2x = 3$, and we know how to solve a congruence such as $2x \equiv 3 \bmod 5$. (We do, don't we? All that's necessary is to realize that $\frac{1}{2}$ makes sense modulo 5—and it does; in fact $\frac{1}{2} \equiv 3$. What that means is that $3 \cdot 2 = 6$ and, since

$$6 \equiv 1 \bmod 5,$$

the number 3 does indeed act the way $\frac{1}{2}$ is supposed to act. To solve the congruence $2x \equiv 3 \bmod 5$ just divide by 2, or, better said, multiply through by $\frac{1}{2}$. The result is $x \equiv \frac{1}{2} \cdot 3 \equiv 3 \cdot 3 = 9 \equiv 4 \bmod 5$, and that checks: $2 \cdot 4 = 8 \equiv 3 \bmod 5$.)

What about the solution of two simultaneous congruences such as

$$x \equiv 5 \bmod 11, \qquad \text{and} \qquad x \equiv 7 \bmod 17?$$

Does that go like the solution of two simultaneous linear equations? No, that's the wrong analogy for sure—in the usual theory there are two equations in two unknowns, whereas here there are two congruences in one unknown—something is fishy. (The solution of two congruences in two unknowns goes pretty much the same way as the usual theory.)

It is clear that simultaneous congruences such as

$$x \equiv a \bmod m \qquad \text{and} \qquad x \equiv b \bmod n$$

don't always have a solution. One obviously bad case is the one in which $m = n$ and the data (a and b) are incompatible: as in

$$x \equiv 4 \bmod 7, \qquad \text{and} \qquad x \equiv 5 \bmod 7.$$

Another bad case (shifting away from prime moduli) is

$$x \equiv 4 \bmod 6, \qquad \text{and} \qquad x \equiv 5 \bmod 10.$$

To see that this is a bad case, observe that every solution of the first of these congruences ($x = 6h+4$) must be even, whereas every solution of the second ($x = 10k + 5$) must be odd. What these bad examples might suggest is the wisdom of restricting attention to the case in which m and n are relatively prime. Is that restriction enough?

Problem 4 C. *If m and n are relatively prime, do the simultaneous congruences*

$$x \equiv a \mod m \qquad \textit{and} \qquad x \equiv b \mod n$$

always have a solution?

4 D. Non–square-free sequences

4 D

A prime number is one that has exactly two distinct factors, namely 1 and itself. (It's understood that we're talking about positive integers.) The closest you can come to satisfying that condition, short of actually satisfying it, is with a number that has exactly three distinct factors. A moment's thought reveals that to satisfy *that* condition a number must be a square; the distinct factors of 9 are 1, 3, and 9. The square of a prime is, from the present point view, as close as any number can be to a prime, short of being a prime.

The prime factorization of a number might contain some primes repeated and others not; in which of those two cases will it have more distinct factors? Look at 20 and 30, for example. The factors of 20 are

$$1,\ 2,\ 4,\ 5,\ 10,\ 20,$$

and the factors of 30 are

$$1,\ 2,\ 3,\ 5,\ 6,\ 10,\ 15,\ 30.$$

Since one of its prime factors, namely 2, enters twice into 20, it ends up having fewer factors. This observation belongs to a general context: a number that is divisible by the square of a prime is likely to have fewer factors than one that is not. A number that is not divisible by the square of any prime is called *square-free*; the observation is that the square-free numbers are the least like the primes, and hence that the non–square-free numbers are the most like the primes.

How often are non–square-free numbers likely to arise? Is that like asking "how often are primes likely to arise?". No—it seems to be the other way around.

Primes are rare. They must be odd, to begin with (the number 2 is exceedingly rare), and therefore there can't be two in a row. Can there be two consecutive odd numbers that are both primes? Sure, lots of times, such as 3, 5 and 11, 13 and 17, 19—and in fact it's a long-standing conjecture that they come that way infinitely often, but no one has ever succeeded in proving that. Can there be three consecutive odd numbers that are all primes? Sure: 3, 5, 7. Any others? No—impossible—of any three consecutive odd numbers at least one is a multiple of 3.

In a similar way square-free numbers are rare. Can there be two, or three, or four consecutive numbers that are all square free? Yes, and yes, and no. For three 5, 6, 7 and 13, 14, 15 are examples, but for four there can be no example: of any four consecutive numbers at least one is a multiple of four, and, therefore, not square-free. It seems, therefore, that despite the plausible considerations that suggested that non–square-free numbers are like primes, the fact seems to be that they are more like composites.

It is a conspicuous and well-known fact that it is quite possible to have two, or three, or four, or even five consecutive numbers that are all composite: examples are 24, 25, 26, 27, 28 and 32, 33, 34, 35, 36. It is even possible to have a hundred consecutive numbers that are all composite, namely

$$101! + 2, 101! + 3, 101! + 4, \ldots, 101! + 101.$$

(The first is divisible by 2, the second by 3, and so on up to the last, which is divisible by 101.) It doesn't take too much imagination to see that the number 100 has nothing to do with the case: there exist arbitrarily long sequences of consecutive numbers that are all composite.

Which way does the corresponding statement about non–square-free numbers go: yea or nay?

Problem 4 D. *What is the largest number of consecutive integers none of which is square-free?*

4 E

4 E. Sums of digits

The "sum of the digits" of a positive integer is just what it sounds like: write the integer in decimal notation and add the digits so obtained. Example: the sum of the digits of 789 is 24, the sum of the digits of 24 is 6, the sum of the digits of 1024 is 7, and the sum of the digits of 2^{20} ($= 1048576$) is 31.

> **Problem 4 E.** *What is the sum of the digits of the sum of the digits of the sum of the digits of* 4444^{4444}?

4 F. Non-divisibility

4 F

Suppose that, temporarily, we agree to regard a finite set of positive integers as good if none of its elements is divisible by another. So, for example, $\{2, 3, 5, 7\}$ is a good subset of the set of integers between 1 and 10 inclusive. Is there a larger one? The question can mean two things. One: can the set as it stands be enlarged and still remain good? Two: is there a good set with more than four elements? A small amount of experimentation reveals that the answer is no to the first question, but yes to the second: the set $\{4, 5, 6, 7, 9\}$ is another good subset of the set of integers between 1 and 10 inclusive. What are the general facts? If a stock of numbers to choose from is prescribed, what's the largest number of elements that a good set can have?

> **Problem 4 F.** *What's the largest number of elements that a set of positive integers between* 1 *and* 100 *inclusive can have if it has the property that none of them is divisible by another*?

Comment. As several times on similar occasions, the number 100 doesn't have much to do with the case, but it helps focus the attention.

4 G. Blockless sequences

4 G

The problem of unending chess had to do with sequences that avoided certain blocks, but the solution left many number-theoretic questions open. The questions are of this kind: how hard is it to avoid blocks?, or, what is the probability of avoiding them?, or, most brutally and most honestly, how many sequences of a specified length succeed in avoiding certain blocks?

Problem 4 G. *How many sequences of* 0*'s and* 1*'s are there, of length* n $(= 1, 2, 3, \ldots)$*, in which there are no blocks of either* 0*'s or* 1*'s of length three?*

Comment. This is the second of an infinite sequence of questions. The first question is the one obtained from the present one by replacing "three" by "two". (Does it make sense to replace "three" by "one"?) That first question is trivial; all it asks is "how many sequences are there in which 0's and 1's always alternate?" Since such a sequence is completely determined by its first term, the answer is obviously two.

4 H

4 H. Maximum product

If for each partition of 10, the product of the parts is formed, how large can that product be? That is: write 10 as the sum of positive integers, and then form the product of the same integers—how large can that be? The numbers involved here are small enough that experimentation is feasible:

$$1+2+3+4 \text{ yields } 24,$$
$$5+5 \text{ yields } 25,$$
$$3+3+3+1 \text{ yields } 27,$$
$$4+4+2 \text{ yields } 32,$$
$$3+3+2+2 \text{ yields } 36,$$

and the latter seems to be unbeatable. What does the answer become when 10 is replaced by 100?

Problem 4 H. *Given a positive integer N, partitioned into positive integers,*

$$N = \sum_{1 \leqq i \leqq k} x_i,$$

what is the maximum value of the product

$$\prod_{1 \leq i \leq k} x_i?$$

4 I. Towers of fives and tens

4 I

The iteration of exponents sometimes leads to curious questions, but before they can be asked the notation must be agreed on. What is a symbol such as 3^{3^3} intended to mean? Is it $(3^3)^3$ or is it $3^{(3^3)}$? In the one case the value would be 27^3 and in the other 3^{27}—and those are very different. The first one is 19,683; the second is a 13-digit number.

The agreement governing the notation is universally accepted—a tower of exponents is always to be read from the top down, and, in particular, 3^{3^3} is always to be interpreted as $3^{(3^3)}$. That makes sense. It would be wasteful to interpret 3^{3^3} as $(3^3)^3$, because the latter, by the elementary laws of exponents, is equal to $3^{3\times 3}$ $(= 3^9)$, for which no other symbol is needed.

Problem 4 I. *Is*

$$10^{5^{10^{5^{10}}}} + 5^{10^{5^{10^5}}}$$

divisible by 11?

4 J. Towers of threes

4 J

How much does the numerical size of an exponential tower depend on the exponents used? It's clear enough that a sequence of powers of 9's such as

$$9, \quad 9^9, \quad 9^{9^9}, \quad 9^{9^{9^9}}, \ldots$$

increases very fast—faster than the corresponding sequence

$$3, \quad 3^3, \quad 3^{3^3}, \quad 3^{3^{3^3}}, \ldots$$

of powers of 3's—but the question is how much faster?

It helps, in speaking of such towers, to have a word that can be used to refer to them and to have a symbol that can be used to indicate them precisely. One good word is *height*: that is the number of levels that enter into the tower. (The heights of the displayed towers of 9's are, in order of appearance, 1, 2, 3, and 4.) A good symbol foı a tower of 9's of height n is $9^{(n)}$. (The displayed towers of 9's are, in order of appearance, $9^{(1)}$, $9^{(2)}$, $9^{(3)}$, and $9^{(4)}$.)

It is clear is that a tower of 3's of height 100, say, is less than (that is, the number it represents is less than) a tower of 9's of the same height; in the notation just introduced

$$3^{(100)} < 9^{(100)}.$$

How much less is it?

It is also true that a tower of 9's of height 100 is less than a tower of 3's of height 200:

$$9^{(100)} < 3^{(200)};$$

to see that, just replace each 9 in the tower by a larger number, namely by 3^3. Can that result be improved?

Problem 4 J. *What is the smallest positive integer n such that $9^{(100)} < 3^{(n)}$?*

4 K

4 K. Irrational punch

Problem 4 K. *Consider a paper punch that can be centered at any point of the plane and that, when operated, removes from the plane all points whose distance from the center is irrational. How many punches are needed to remove every point?*

CHAPTER 5

GEOMETRY

Introduction

The problems in this section are about geometry, in the sense that they concern standard geometric concepts (such as straight lines, triangles, squares, distances, congruence). They are, however, not geometry in the sense of Euclidean constructions, and they are not geometry in the sense of the algebraic manipulations that frequently enter the subject called "analytic geometry". The focus will often be on the sort of thing that analysts think about, such as the minima involved in finding shortest distances; the methods of attack will, however, usually be geometric.

5 A. Pyramid and tetrahedron

5 A

Problem 5 A. *Suppose that P is a pyramid (equilateral triangles on a square base) and T is a regular tetrahedron (with faces congruent to the triangular faces of P). If P and T are joined by gluing one of the faces of T onto one of the triangular faces of P, how many faces does the resulting surface have?*

5 B

5 B. Join closed finite sets

Call a subset of the plane *join closed* if the straight line joining any two of its points contains at least one other point of the set.

The definition looks disarmingly simple, but an attempt to construct non-trivial examples is likely to lead to quick discouragement. One trivial example is the plane itself, and another is a straight line, any straight line, in the plane. These examples satisfy the condition of the definition, but they are big—they are infinite.

Are there any finite examples? Sure—just consider a set consisting of three points on a line, or, for that matter, of any finite number of points on a line—but such examples satisfy the condition of the definition so easily, that most people would probably declare them to be trivial also. For the time being, therefore, a non-trivial example might be defined as one that is "small" and not linear. Are there any?

Problem 5 B. *If a finite set is join closed does it have to be a subset of a line?*

5 C

5 C. Maximal triangle

Problem 5 C. *For each square of area* 1*, what's the largest area that a triangle inside it can have?*

5 D

5 D. Minimal triangle

Problem 5 D. *For each point in the first quadrant, which is the line through that point that, together with the positive x and y axes, bounds the triangle of minimal area?*

5 E

5 E. Minimal roads

Problem 5 E. *There is a house at each of the four corners of a square of side length* 1 *mile. What is the shortest road system that enables each of the inhabitants to visit any of his neighbors?*

5 F. Short straightedge

5 F

The classical Greek constructions with straightedge and compass are these: given two points, construct the (infinite) straight line through them, and given two points, construct the circle through one with center at the other. Every other permitted geometric construction follows from these two. Some familiar examples: divide a given segment into seven segments of equal lengths, given a segment and a triangle with a specified edge, construct a similar triangle with the specified edge replaced by the given segment; given a straight line and a point not on it, construct a parallel line through the point.

Suppose now that the rules remain the same but the tools are restricted: the straightedge is one inch long, and the compasses can open to a radius of not more than one inch. Can the Greek constructions still be performed?

> **Problem 5 F.** *Given two points in the plane a mile apart, can they be joined with a straight line segment* (*by the Greek rules*), *using a short straightedge and a rusty compass*?

5 G. Disjoint triangles

5 G

Do three points in the plane determine a triangle? The answer is: usually. The exceptional case occurs if they all lie on a line (if they are collinear)—in that case the triangle degenerates to a segment and doesn't deserve to be called a triangle. Similarly, six points in the plane determine two triangles—usually. If they are collinear, there is obviously something wrong; what if just some three of them are collinear—is that bad? No, not necessarily (see Figure 3): and even if four of them are collinear, it is possible for them to be the vertices of two disjoint trian-

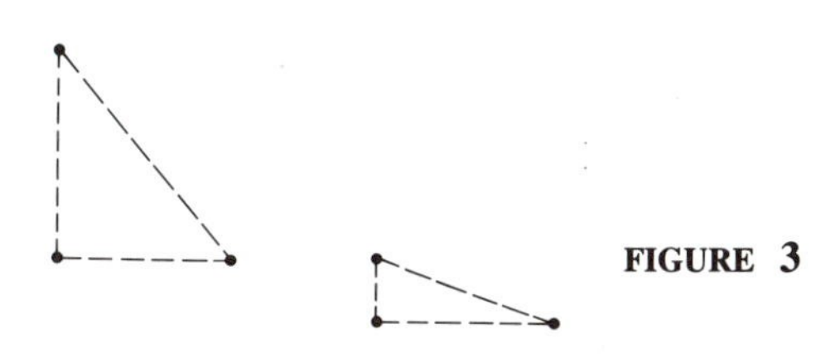

FIGURE 3

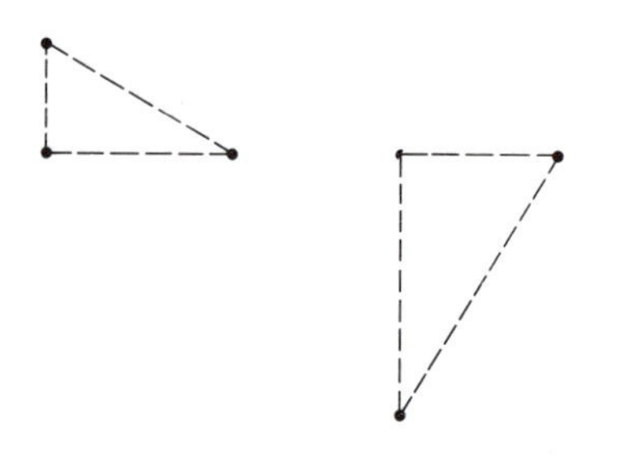

FIGURE 4

gles (see Figure 4). If, however, five are collinear, or all six are, then the position becomes degenerate again.

There is a different kind of distressing situation that can arise with six points even if no four or five of them are collinear; in Figure 5 that is the case, but, nevertheless, there do not exist two disjoint triangles that have the six points in question as their vertices.

FIGURE 5

To avoid all possible misunderstanding, a "triangle" is defined here so as to include both boundary and interior, and two triangles are called "disjoint" if they have no points in common at all, not even boundary points. In order to be sure that a prescribed bunch of points can be the vertices of a bunch of disjoint triangles, some condition restricting collinearity is obviously necessary. It is, however, not obvious that even the strongest possible restriction is sufficient; it could be that something other than partial collinearity could be an obstruction.

Problem 5 G. *If no three of a prescribed set of* 3000 *points in the plane are collinear, does it follow that they are the vertices of a thousand disjoint triangles*?

5 H. Maximal rectangle

5 H

Problem 5 H. *For an ellipse with major axis twice as long as the minor axis, what is the ratio of the area of the ellipse to the area of the largest inscribed rectangle?*

5 I. Bisection of triangles

5 I

What's the most efficient way to bisect a triangle? The question doesn't quite make sense; "efficient" and "bisect" must be defined. Very well: let "bisect" mean bisect the area, and call one bisection more "efficient" than another if the length of the curve it uses is shorter. Suppose, for instance, that we are given an isosceles right triangle; Figure 6 shows four possible ways to bisect it. The most of efficient of these is pretty clearly the one on the left.

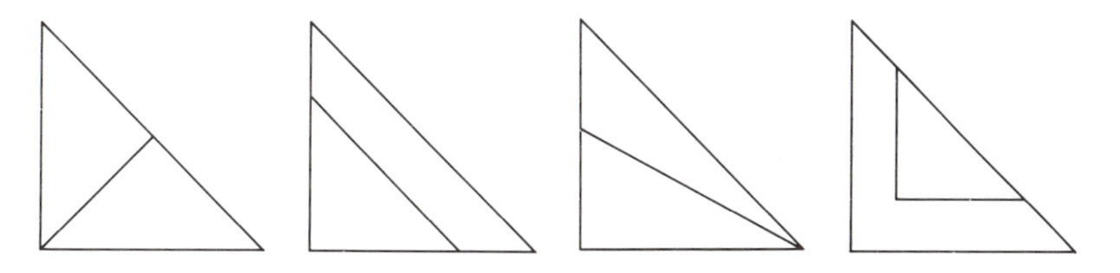

FIGURE 6

Problem 5 I. *What's the most efficient way to bisect an equilateral triangle?*

5 J. Cavalieri congruence

5 J

Two triangles T and T' in the plane are *Cavalieri congruent* if they can be positioned (by translation and rotation) so that for each horizontal line L the intersections $L \cap T$ and $L \cap T'$ have the same length. ("Triangle" here means interior and perimeter; generalizations to other sets suggest themselves naturally. As for directions, they all play the same role; "horizontal" is just a convenience.) It is a classical fact, and an elementary one, that if two triangles are Cavalieri congruent, then they have the same area; just integrate. Could the converse possibly be true—perhaps with some additional conditions?

Problem 5 J. *If two triangles have the same area, are they necessarily Cavalieri congruent?*

5 K

5 K. Extreme points, two-dimensional

A set (in the plane, or for that matter in a line or in three-dimensional Euclidean space) is called convex if it looks like a cracker, not like a pretzel. Formally: convexity requires that the line segment joining any two points of the set lie entirely in the set. Examples: a single point, an interval, a line, or, in the plane, the interior of a triangle, or of any circular disk. Examples in three-dimensional space: a solid ball, a solid cone. Observe that a convex set may or may not be bounded, and it may or may not be closed.

Of interest in the study of a convex set are the extreme points of the set. Extreme point means omittable point—a point of the set with the property that if we discard it from the set, the remainder is still convex. Example: if we discard an interior point from an interval, the result is not convex—the interior point was not an extreme point. On the other hand if we omit one of the end points of a closed interval, the result is still convex—an end point is an extreme point. An open interval has no extreme points at all. A half open interval has exactly one, and so does a closed half infinite ray.

In two and three dimensions examples get more interesting. What are the extreme points of a closed triangle? Answer: the three vertices. What about a closed square with a closed semi-disk sitting on its top edge, such as the one shown in Figure 7? Answer: the semi-perimeter and the bottom two vertices. What about a closed disk? Answer: the entire perimeter. What about an open disk? Answer: none. What about a planar ice cream cone, such as the one shown in Figure 8. Answer: the bottom point and the top perimeter. What about a closed ball? Answer: the entire shell. What about an infinite ice cream cone? Answer: just the bottom vertex. What about the infinite ice cream cone without the bottom vertex? Answer: none. What about an entire line in the plane, or an entire plane in space? Answer: none. Conclusion from these examples:

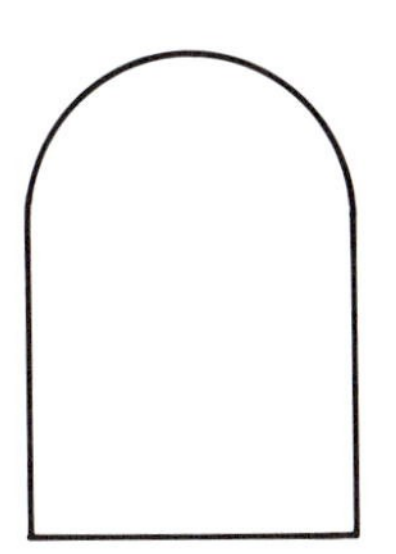

FIGURE 7

FIGURE 8

if a set is not closed, it may not have any extreme points at all; if a set is not bounded, it may not have any extreme points at all.

Take a weird example: an open disk, together with a random bad subset of its perimeter. Extreme points: just those bad points. Conclusion: the set of extreme points can be a bad set. Conjecturable reason: the set was bad to start with. Question: what about the set of extreme points of a closed convex set?

Problem 5 K. *Is the set of extreme points of a closed convex set in the plane always closed?*

Comment. The examples above included sets that have no extreme points at all—their set of extreme points is empty. That's just fine: the set of extreme points in that case is closed.

5 L

5 L. Extreme points, three-dimensional

Sometimes the passage from the plane to space is automatic, and sometimes it is surprising. Which is it this time?

Problem 5 L. *Is the set of extreme points of a closed convex set in three-dimensional space always closed? What if the set is compact (closed and bounded)?*

5 M

5 M. Extreme points, finite-dimensional

All this talk about extreme points may be well and good—but do they have to exist? Sure, there are lots of examples of extreme points—but do extreme points have to exist for every compact convex set? The answer is obviously no if the given set is empty—are there other bad cases that must be excluded?

Problem 5 M. *Does every non-empty compact convex set in a finite-dimensional Euclidean space have at least one extreme point?*

Comment. Some of the examples seen above show that closed (but not necessarily compact) convex sets are not good enough: consider a line, or, in the plane, the closed band between two parallel lines.

5 N

5 N. Closed convex hulls

If a set is not convex, it can be made convex. The point is that for quite an arbitrary set S, there always exist convex sets that cover S—if worse comes to worst the entire space is one of them. It could well be, sometimes, that that's the only one. (Example: let S consist of the x and y axes in the plane.) If there are many convex sets that cover S, the intersection of any two of them is another one, and the same is true of the intersection of three, or a million, or an infinite number of them, if there are that many. In fact the intersection of *all* the convex sets that cover S is a convex set that covers S, and, obviously, it is the smallest such convex set. (Clear? By its very construction, it is included in every other one.) That intersection, that smallest convex set that covers S, is called the convex hull of S.

The convex hull of a set doesn't have to be closed. (Trivial example: just take a set that is already convex, but just happens not to be closed; its convex hull is itself.) Sometimes it is useful to form the so-called closed convex hull of S—that is the smallest closed convex set that covers S. To form it, just consider the closed convex sets that cover S and intersect them all. If we start with a closed set, can the two different constructions lead to two different answers?

Problem 5 N. *Is the convex hull of a closed set necessarily closed?*

Comment. The concepts of convex hull and closed convex hull play a fundamental role throughout the theory of convex sets. They enter, in particular, the Krein–Milman theorem (Solution 5 M): a great improvement of that theorem is the assertion that not only do compact convex sets have extreme points, but, in fact, they have "enough" of them. Explanation: each such set is the closed convex hull of the set of its extreme points.

CHAPTER 6

TILING

6 A. Chomp, square

6 A

Consider a rectangle subdivided into squares, such as, for instance, the 5×9 rectangle in Figure 9. Chomp is a two-person game played on such

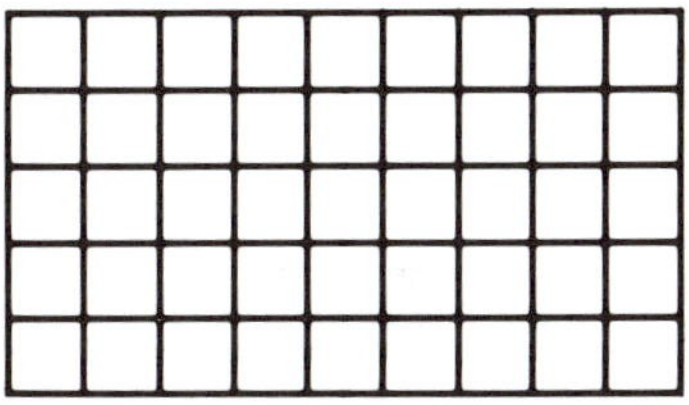

FIGURE 9

a board, as follows. The first player selects any one of the 45 squares, say for instance the one marked "1" in Figure 10 and removes it from the board together with all the squares above and to the right of it. (One interpretation of the game has a piece of chocolate on each square; the player "removes squares" by eating the chocolates on them. Alternatively, the board is made of chocolate, and a move is a bite, a "chomp",

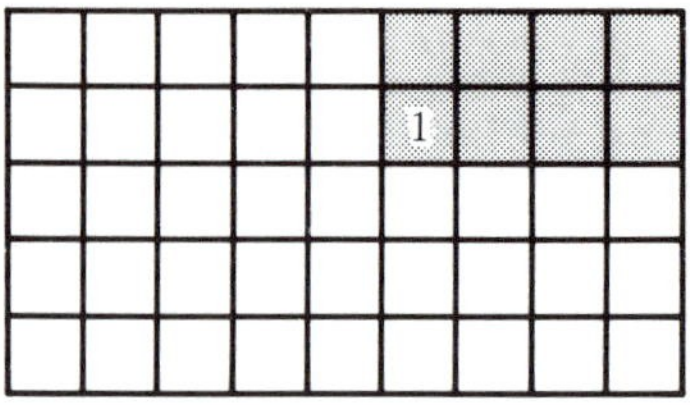

FIGURE 10

that removes a northeast corner of the board.) The second player, in turn, selects a square, any of the squares not yet removed, say, for instance the one marked "2", and removes from the board the square just

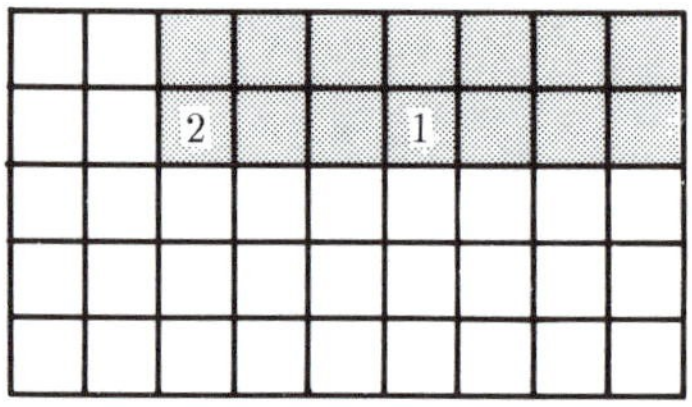

FIGURE 11

selected and all the squares above and to the right of it. Next, it's the first player's turn again, to select a square, any of the squares not yet removed, say for instance the one marked "3", and to remove from the

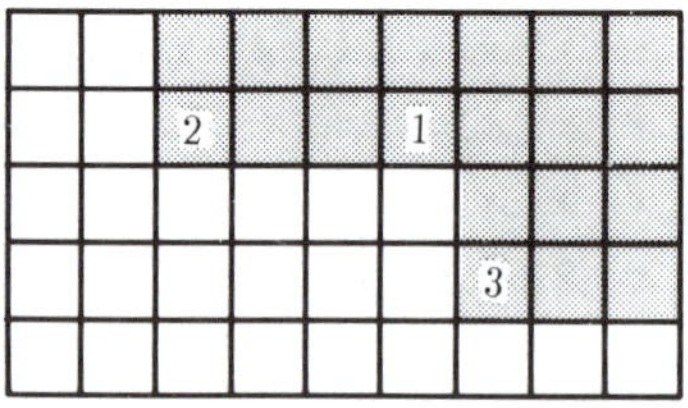

FIGURE 12

board the square just selected and all the squares above and to the right of it. The game continues this way till it has to end—which is when one player removes the bottom left square, and, by doing so, *loses* the game. (In the story version, the bottom left corner is poisoned, and its consumption premanently removes the player who removed it.)

Problem 6 A. *If the chomp board is a square, can either player force a win?*

6 B. Chomp, thin

6 B

Some rectangles are "thin" and some are not; the least thin, in the sense intended, are the squares. Chomp on square boards has been settled; what happens with thin boards? If the board is 1×1, the first player loses just because he has to move; if the board is $1 \times n$, for $n > 1$, the first player can force a win simply by removing all the squares except the bottom left. The first case that requires a little thought is $2 \times n$ ($n \geqq 2$).

Problem 6 B. *If the chomp board is* $2 \times n$, $n = 2, 3, 4, \ldots$, *can either player force a win?*

6 C. Chomp, infinite

6 C

The definition of chomp makes sense even on infinite boards, so long as it makes sense to speak of the bottom left corner. It makes sense, in particular, for boards of the shape $2 \times \infty$—boards, that is, with two rows that stretch out toward the right ad infinitum.

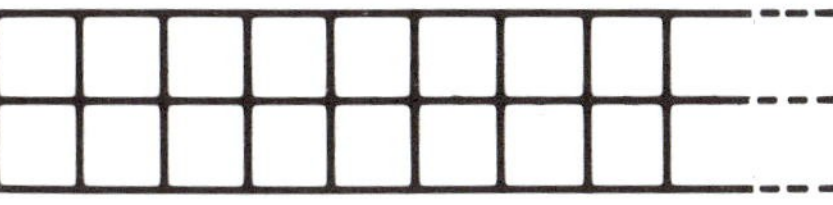

FIGURE 13

Problem 6 C. *If the chomp board is* $2 \times \infty$, *can either player force a win?*

6 D

6 D. Chomp, finite

The preceding chomp problems all asked about the possibility of forced wins. The infinite and finite cases turned out to be different, and surely that is not too shocking. In the finite cases so far concerned, fat squares and thin rectangles, the problem has been settled in what seem to be two diametrically different cases. Can the general case be settled similarly?

> **Problem 6 D.** *If the chomp board is finite, can either the first or the second player always force a win, and, if so, which?*

6 E

6 E. Plane, three colors

> **Problem 6 E.** *If every point of the plane is colored one of three colors, are there necessarily at least two points of the same color exactly one inch apart?*

6 F

6 F. Plane, ten colors

> **Problem 6 F.** *If every point of the plane is colored one of ten colors, are there necessarily at least two points of the same color exactly one inch apart?*

Comment. This is word for word the same problem as 6 E with just one exception: "three" was changed to "ten".

6 G

6 G. Domino tiling

Form a 10×10 square subdivided into a hundred 1×1 squares, and define a domino to be a 1×2 rectangle consisting of two small squares sharing a side (as in Figure 14). The total area of fifty dominoes is 100, the same as the area of the large square, and, of course, the large square can be tiled with fifty dominoes.

If the two squares at the lower left are omitted from the large square (as in Figure 15), the remaining area is 98, which can be tiled with 49 dominoes.

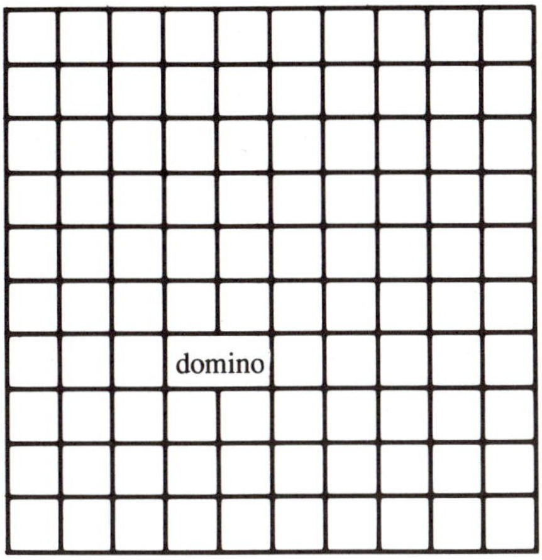

domino

FIGURE 14

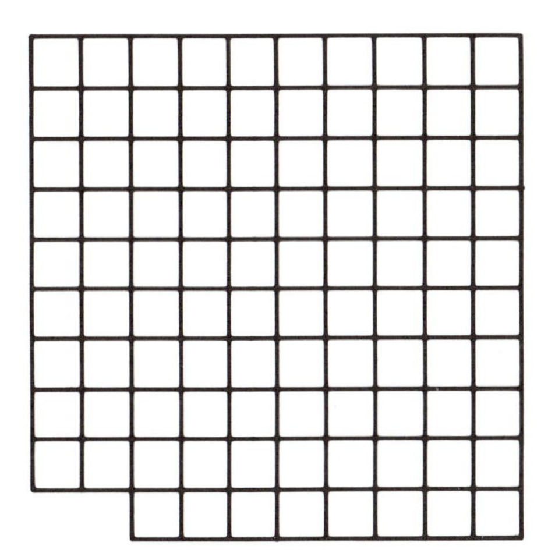

FIGURE 15

If, instead, the two corner squares of the bottom row are omitted from the large square (as in Figure 16), leaving an area of 98 again, can that remainder be tiled with 49 dominoes? Sure, why not: just fill in the eight squares between the two omitted corners with four dominoes, and tile the remaining 90 in any obvious way.

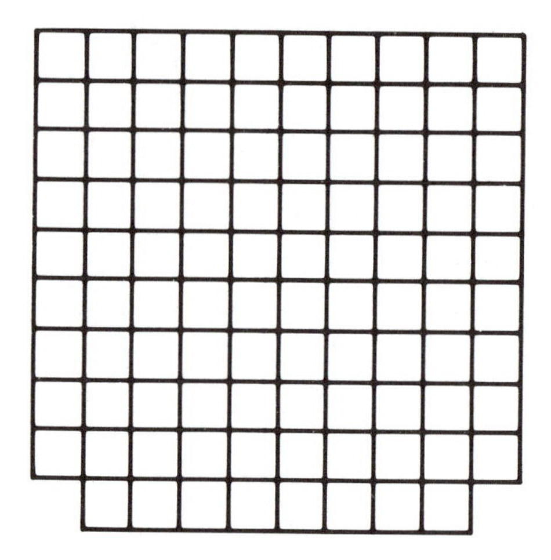

FIGURE 16

Problem 6 G. *If two diagonally opposite corner squares are removed from a* 10×10 *square subdivided into a hundred* 1×1 *squares (as in Figure* 17*), can the remainder be tiled with* 49 *dominoes?*

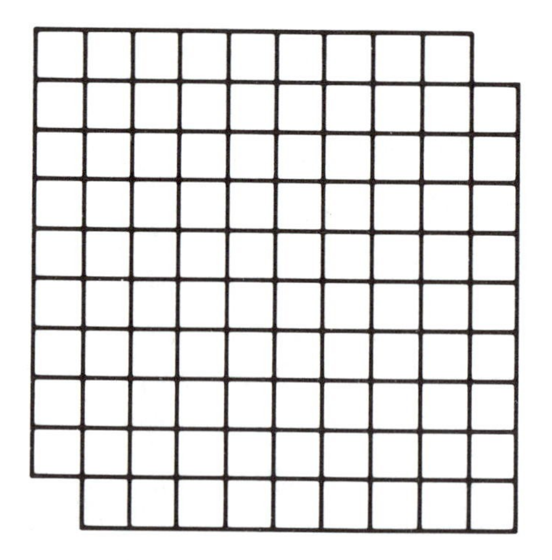

FIGURE 17

6 H. Domino tiling, other

6 H

What happens to the tiling problem if, in accordance with the suggestion of the hint for Problem 6 G, the number 10 is replaced by 8 (usually a more convenient number), and, in accordance with the implied suggestion of Solution 6 G, two squares of different colors are removed from the checkerboard? The method breaks down; does the result break down too?

Problem 6 H. *Delete two squares of different colors from a checkerboard. If a domino is defined to be a* 1×2 *rectangle consisting of two squares (of the same size as the ones on the checkerboard) sharing a side, then the total area of* 31 *dominoes is* 62*—the same as the area of the checkerboard after the deletion. Can that area be tiled with* 31 *dominoes?*

6 I. Dominoes, long

6 I

The domino tiling problem can be generalized in various ways, some more obvious than others. The most obvious generalization is to replace 10 by any other positive integer n. For some n's the answer to the generalized question is clearly no: if, for instance, n is odd, then the removal of two corner squares still leaves an odd number of small squares, and anything tiled by dominoes must have an even number. In general the possibility of tiling after deletions depends on the relation between the size and shape of the originally given board and the size and shape of the deleted material.

Consider, as an example of the sort of thing that can happen, the checkerboard ($n = 8$) and the tiling problem with dominoes that are 1×3 rectangles. In this case the number of remaining squares after deletion must be divisible by 3; the easiest number to delete to achieve that end is 1.

Problem 6 I. *Delete any one square from a checkerboard. If a domino is defined to be a* 1×3 *rectangle consisting of three squares (of the same size as the ones on the checkerboard), then the total area of* 21 *dominoes is* 63*—the same as the area of the checkerboard after the deletion. Can that area always be tiled with* 21 *dominoes, or never, or does it depend on which square is deleted?*

6 J

6 J. Dominoes, corner

Many other variations on the tiling theme are possible. Here is just one more, having to do with tiles that are not rectangles.

Problem 6 J. *Delete any one square from a checkerboard. If a corner (angle-iron) is defined to be an L-shaped figure consisting of three squares (of the same size as the ones on the checkerboard), then the total area of* 21 *corners is* 63*—the same as the area of the checkerboard after the deletion. Can that area always be tiled with* 21 *corners, or never, or does it depend on which square is deleted?*

FIGURE 18

6 K

6 K. Integer rectangles

Consider a rectangle, tiled with, say, five rectangles, something like in Figure 19. Assume that each of the small rectangles, the tiles, is semi-

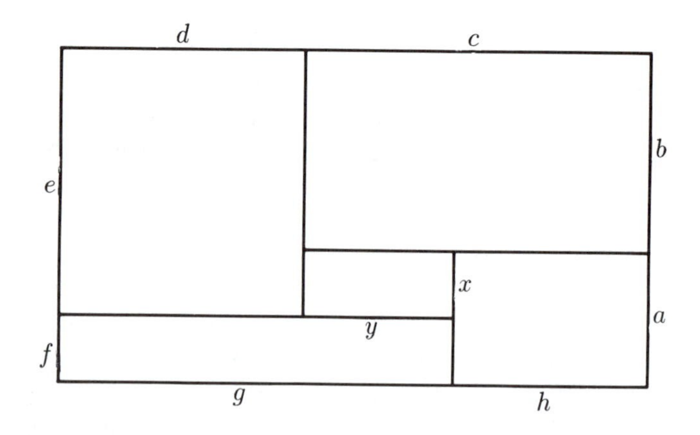

FIGURE 19

integral, meaning that the length of at least one side of each tile is an integer. Does it follow that the given large rectangle is semi-integral?

It is convenient to argue by contradiction: assume that the large rectangle is not semi-integral, and see what follows. It is known that either a or h is an integer, and a and h play essentially symmetric roles; assume, with no loss of generality, that a is an integer. Consequence (since each tile is semi-integral): c is an integer. These two steps can be repeated over and over again: since c is an integer, d is not, therefore e is, therefore f is not, therefore g is, therefore h is not. So far the argument made use of the assumptions about each of the four outside tiles. Observe now that the height x of the inside tile is equal to $a - f$, and its base y is $g - d$, neither of which can be an integer. A contradiction has arrived; the large rectangle must be semi-integral. Is the conclusion an accidental property of the number 5 (the number of tiles here used) and of the particular way the tiles are laid out?

> **Problem 6 K.** *Can a non–semi-integral rectangle be tiled with rectangles that are semi-integral?*

Comment. The argument used above can be repeated, in simpler forms, for tilings that use only one tile (a silly case), or two, or three, or four. For tilings with six tiles, it gets complicated; for tilings with a hundred, it gets hopeless.

CHAPTER 7

PROBABILITY

7 A. Loaded dice

7 A

Amateurs of probability theory are sometimes befuddled by the way the probabilities on a pair of dice turn out, and object to it. Why should it be that the probability of rolling a sum of 2 is $\frac{1}{36}$, but the probability of rolling 4 is $\frac{1}{12}$ $(= \frac{3}{36})$—why are all those numbers different and confusing? Why couldn't things be arranged so that all the probabilities, the probabilities of all possible sums from 2 to 12 inclusive, are the same?

> **Problem 7 A.** *Is it possible to load a pair of dice so that the probability of occurrence of every sum from* 2 *to* 12 *shall be the same?*

Comment. Since there are eleven possible sums, the question can be phrased this way: is it possible to load a pair of dice so that the probability of the occurrence of each of the sums $2, 3, \dots, 12$ shall be $\frac{1}{11}$?

The question has a mathematical interpretation and a physical one. The mathematics asks whether it is possible to assign probabilities $p_1, p_2, p_3, p_4, p_5, p_6$ to one die and probabilities $q_1, q_2, q_3, q_4, q_5, q_6$ to the other so that when the probabilities of $2, 3, \dots, 12$ are calculated all the results are equal to $\frac{1}{11}$. The physics asks whether, granted that the answer to the mathematics question is yes, is it actually possible to design the distribution of weights, and then to pour the plastic (ivory?) that dice are made of, so as to achieve those probabilities. The present question is the mathematical one.

7 B

7 B. Honestly loaded dice

Is there a dishonest way of loading honest dice?

> **Problem 7 B.** *What are all the possible ways of loading a pair of dice in such a way that the probability of the occurrence of each sum from* 2 *to* 12 *is the same as it is for honest dice?*

7 C

7 C. Renumbered dice

If we change the probabilities of a pair of dice, we'll not end up with a pair that yields the same sum probabilities as the "classical" ones—that's what Solution 7 B says. What if we do not change the probabilities, but change the faces?

> **Problem 7 C.** *Are there ways of changing the numbers of dots on the faces of a pair of honest dice in such a way that when they are rolled simultaneously the possible sums are the same as always, from* 2 *to* 12, *and, at the same time, the probability of the occurrence of each sum from* 2 *to* 12 *is the same as it is for ordinary dice?*

7 D

7 D. Transitive probability

Is it possible that most people find apple pie preferable to blueberry, most people find blueberry pie preferable to cherry, and, at the same time, most people find cherry pie preferable to apple? The paradoxical answer is yes, and after a little thought even the paradox goes away, even in very small populations. Think, for instance, of three people, 1, 2, and 3, whose orders of preferences are indicated in the rows:

$$\begin{array}{cccc} 1 & A & B & C \\ 2 & B & C & A \\ 3 & C & A & B. \end{array}$$

Clear? Two out of three (namely 1 and 3) prefer A to B, two out of three (1 and 2) prefer B to C, and two out of three (2 and 3) prefer C to A. Nothing to it.

The situation described above is deterministic; can the same thing happen probabilistically? A standard old question concerns three students preparing for an examination. Is it possible that A will probably get a higher grade than B, B will probably get a higher grade than C, and, at the same time, C will probably get a higher grade than A? How is "probably" to be interpreted in questions such as this? One reasonable interpretation is "with probability more than $\frac{1}{2}$".

In slightly more dignified language the probability question can be phrased this way: do there exist three random variables f, g, and h such that each of the three numbers

$$\text{Prob}\,(f > g), \quad \text{Prob}\,(g > h), \quad \text{Prob}\,(h > f)$$

is greater than $\frac{1}{2}$? A short period of cogitation should show that the answer is yes.

In a related but slightly more suggestive context, suppose that there are three different examinations of which one is to be chosen at random (each having probability $\frac{1}{3}$). It is then perfectly feasible that on the first the candidates will rank ABC, on the second BCA, and on the third CAB. If that is the case, then, indeed,

$$\text{Prob}\,(A > B) = \text{Prob}\,(B > C) = \text{Prob}\,(C > A) = \frac{2}{3}.$$

What these examples show is that certain relations that are intuitively thought to be transitive are in fact not transitive. Can such non-transitive phenomena occur in the presence of independence?

Problem 7 D. *Is it possible to load each of three dice so that when they are rolled, the first will probably show a higher number than the second, the second will probably show a higher number than the third, and the third will probably show a higher number than the first?*

7 E. Best horse loses

7 E

What's a good way to compare the performance of, say, two racehorses? One way, perhaps, is by time trials: make the horses race a fixed standard distance over and over again and keep track of the time it takes each of them to do so. A statistical summary of the information so gathered will answer (or at least approximately answer) questions such as these: what is the probability that horse X will cover the distance in a time not greater than some specified t, and how does that compare with the answer to the same question for horse Y? The resulting distribution functions (considered as functions of t) look as if they would give the information usually wanted. If, for instance, no matter what t is, X has a higher probability of running the race in less time than t than Y does, then X is the horse you want to bet on. Right?

Problem 7 E. *If, for each time t, the probability that racehorse X covers a track in time not greater than t is greater than the corresponding probability for racehorse Y, does it follow that the probability of X beating Y in a race is at least $\frac{1}{2}$?*

7 F

7 F. Odd man out

The simplest two-person gambling game is to toss a coin: heads you win, tails you lose. An alternative and frequently used version is "matching": you and I each toss a coin; you win if our coins match (both show heads or both show tails), and you lose if they don't match. The latter version is easy to generalize to the problem of randomly selecting one out of three people to be the winner. Each of the three of us tosses a coin, and if the coins fall one head and two tails, or one tail and two heads, then the odd one wins; if they fall all heads or all tails, then nobody wins, and the toss is repeated. Once we get started on this road, it is trivial to continue forever: the random selection of a winner by "odd man out" (a cliché phrase that seems to apply here very neatly) works just as well for four, five, or any number of people. The question is: how well is that?

Problem 7 F. *If n people toss for a winner by "odd man out", $n = 1, 2, 3, \ldots$, what is the probability that the issue is settled on the first toss?*

7 G. Odd one of four

Problem 7 G. *If four of us toss for a winner by "odd man out", what is the probability that I win?*

7 H. Probability $\frac{1}{3}$

Problem 7 H. *Is there a gambling game (with an honest coin) for two players, in which the probability of one of them winning is $\frac{1}{3}$?*

Comment. What's the probability that when five coins are tossed (one after another or simultaneously, from the present point of view it doesn't matter which) they break four to one: four heads and one tail or else four tails and one head? That was exactly the "odd man out" question, and the answer turned out to be $\frac{10}{32}$. Never mind what the answer is; the present emphasis is on the "shape" of the answer. Does it make sense that the answers to all questions of this sort seem to be fractions whose denominator is a power of 2? Assuming (the assumption has been silently made till now, and is hereby explicitly being put into force) that the coins are honest—the probability of head and the probability of tail are both equal to $\frac{1}{2}$—even the briefest exposure to probability calculations indicates that all answers to questions about coins are indeed fractions of that kind. The present question, therefore, is intended to be a challenge: doesn't it look as if the answer ought to be no?

7 I. Probability Answers

Problem 7 I. *Is there a gambling game (with an honest coin) for two players, in which the probability of one of them winning is $\frac{1}{\pi}$?*

7 J

7 J. Balls in boxes

Problem 7 J. *A hundred balls, fifty black and fifty white, are distributed between two boxes, a box is selected at random, and from it a ball is drawn at random. Does the probability of getting a white ball this way depend on how the balls were distributed?*

7 K

7 K. Expectation of sums

Take a yardstick, break it in two at some random place, keep the part in your left hand, and throw away the part in your right hand. Do it again: new yardstick, new break, add the kept left-hand part to the previously kept one, and discard the right-hand part. How long will it be before the parts that you kept add up to at least one yard?

Problem 7 K. *Choose one after another a sequence of numbers in the interval* $[0, 1]$ *(uniform probability distribution); what is the expectation of the (smallest) number of numbers needed for the sum of the chosen numbers to be* 1 *or more?*

7 L

7 L. Numbers in a hat

There are a hundred pieces of paper in a hat with a number written on each. Caution: the numbers need not be the first hundred positive integers, they needn't be integers at all, they needn't be positive—they are totally unknown real numbers, which may be 3, or -17, or e^{π}, or $1{,}000{,}000^{\sqrt{2}}$. All that is assumed is that the hundred numbers are all different—no coincidences. A gambling game is to be played as follows. You draw the slips of paper out of the hat, blindly, at random, look at each as you draw it, and then discard it forever before drawing the next. You can stop drawing at any time: you can stop after the first draw, or any other; if you haven't stopped with any of the first 99, you must stop with the 100th. You win if the number you stop with is the largest number in the hat—the largest ever—larger than any that have been discarded and larger than any that remain; in any other case you lose. The ante that you have paid the casino to be allowed to play the game is, say, $1.00. How much payoff should you receive when you win in order to make the game fair?

Problem 7 L. *If the payoff in the "numbers in a hat" game is* \$5.00 *for winning, is the expectation of the player positive or negative?*

CHAPTER 8

ANALYSIS

8 A. Towers of roots

8 A

Towers of exponents are not uncommon things—towers of fives and towers of threes have been seen, for instance, in Problems 4 I and 4 J. Such towers are, however, finite; can infinite ones make sense?

Problem 8 A. *What can something like*

$$\sqrt{2}^{\sqrt{2}^{\sqrt{2}^{\sqrt{2}^{\cdot^{\cdot^{\cdot}}}}}}$$

possibly mean, and how much is it?

8 B. The L^1 norm of a derivative

8 B

What can be said about the values of a function when something is known about the values of its derivative? In order for the question to make sense, we must be sure, first of all, that the function has a derivative, and, equally important, that the usual calculus relation between a function and its derivative is applicable. "The usual calculus relation" refers to the possibility of recapturing a function from its derivative (by integration)—a relation that is not guaranteed by the mere existence of the derivative. (The technical term for functions that are well behaved in that sense is "absolutely continuous".)

Suppose then that the usual calculus relation holds, and suppose that we are told of a function on the unit interval that starts and ends at 0 and that does not grow (or shrink) too much too often. That's vague, of course; how can such a restriction be made precise? To say that the function doesn't grow too much might be interpreted to mean that its derivative keeps from being positive and large; similarly to say that the function doesn't shrink too much might mean that its derivative keeps from being negative and large (in absolute value). To say that the derivative doesn't do these things "too often" might mean that it's allowed to do them, that it's allowed to have a derivative of $+100$ sometimes and a derivative of -100 sometimes, but that "on the average" the absolute value of the derivative stays under control. The phrase "on the average" almost always refers to a mean value—an integral.

Problem 8 B. *If f is an absolutely continuous function on $[0, 1]$ such that $f(0) = f(1) = 0$, and if*

$$\int_0^1 |f'(x)|\, dx = 1,$$

what can be said about $f(\frac{1}{2})$? For example, is it possible that $f(\frac{1}{2}) = 0$?, or 1?, or $\frac{1}{2}$?, or -1?, or 2?

Comment. It seems intuitively plausible that it is *not* possible that $f(\frac{1}{2}) = 100$. The idea is that if f is going to go from 0 to 100, and back again, it has to do a lot of growing and then a lot of shrinking—its derivative has to have quite large absolute value quite often—and that's exactly what the condition on the integral of the derivative seems to rule out. Incidentally, the expression

$$\int_0^1 |f'(x)|\, dx$$

occurs often in functional analysis and is called the L^1 norm of the derivative f'; the usual symbol for it is $\|f'\|$ or $\|f'\|_1$.

8 C. The boundedness of a derivative

8 C

What can be said about the values of the derivative of a function when something is known about the function itself? It's a trivial task to find a function that takes arbitrarily large values with a derivative that remains bounded: consider, for instance, $f(x) = x$. It is, however, more usual to expect that if the function becomes large, then so does its derivative: consider, for instance, $f(x) = x^2$. One difference between the two examples is that the first is uniformly continuous, but the second is not. Is that the main difference?

> **Problem 8 C.** *If a function f is continuously differentiable and uniformly continuous, does it follow that the derivative f' is bounded?*

Comment. If the derivative is bounded, then the function is uniformly continuous; after a brief meditation on the mean-value theorem that statement becomes obvious. The present question is whether the converse is true.

8 D. Log square integrability

8 D

The techniques of calculus often answer questions that a calculus course rarely asks. Example: for which exponents p does the equation

$$f(x) = x^p$$

define an integrable function f in the unit interval? (Since the end points, especially 0, are often troublesome, in this context it is safest to consider only the open unit interval $(0,1)$.) The way to find the answer is to try to integrate and see what happens; the result is that all goes well if $p > -1$, and all goes badly otherwise. Other functions can be trickier. Example:

$$f(x) = \log x.$$

Is it integrable? In other words: does

$$\int_0^1 \log x \, dx$$

make good sense? Calculus still gives the answer, and the answer is yes; we are looking at an "improper" but perfectly healthy integral. Its value is -1.

Problem 8 D. *Is the logarithm function square integrable in the unit interval?*

Comment. What the question amounts to is this: is the integral

$$\int_0^1 |\log x|^2\, dx$$

finite or not? In the language of functional analysis: does the function log belong to $L^2(0,1)$ or not?

8 E

8 E. Log over square integrability

Problem 8 E. *What's the value of*

$$\int_0^\infty \frac{\log x}{1+x^2}\, dx?$$

Comment. The main question, presumably, is whether or not the integrand is integrable. If the answer is yes, then, presumably, it can be evaluated, but the question makes some sense even if the answer is no: in that case the answer might be $+\infty$ or $-\infty$.

8 F

8 F. Universal chords

Look at the graph of a continuous function f defined on the closed unit interval $[0,1]$ that vanishes at both ends (Figure 20). Call a positive number c a *chord* of the function f if there exists a number x such that both x and $x+c$ are in $[0,1]$ and $f(x+c) = f(x)$. Example: the number $c = 1$ is a chord of f, because, by assumption, $f(0) = f(0+1)$, so that the conditions are satisfied by $x = 0$ and $c = 1$. Figure 21 indicates a less trivial possibility: the horizontal segment joining the points $(x, f(x))$ and $(x+c, f(x+c))$ has both its end points on the graph, and

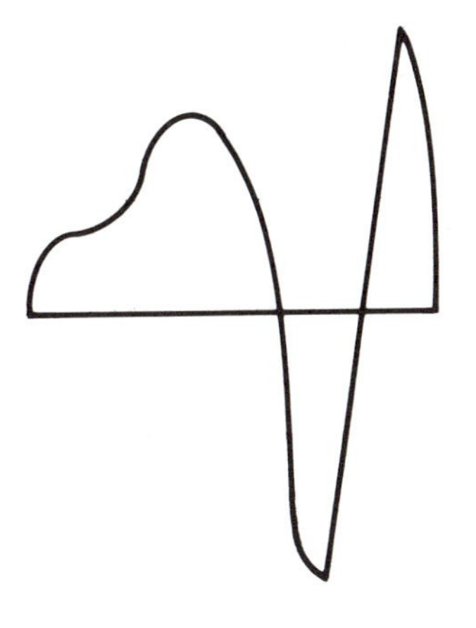

FIGURE 20

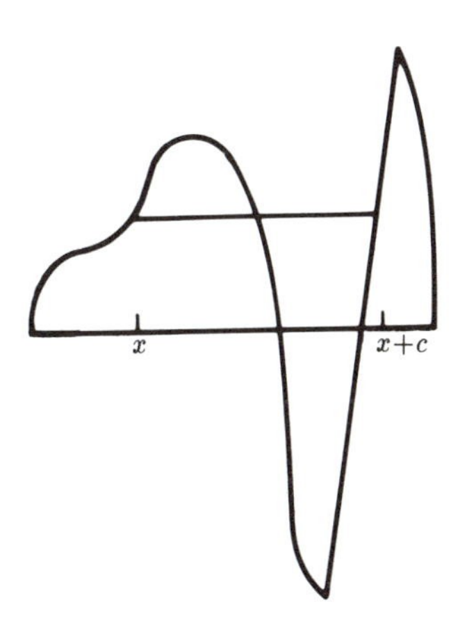

FIGURE 21

its length is, by construction, equal to c; for that reason the number c is called a chord of f.

A number is called a *universal chord* in case it is a chord of every function of the kind under consideration (that is, every continuous function f on $[0, 1]$ with the property that $f(0) = f(1)$). Example: the number $c = 1$ is a universal chord. What other numbers are universal chords?

Problem 8 F. *Is $\frac{1}{2}$ a universal chord? What about .1? What about .2? What about .3? What about $\frac{1}{e}$?*

8 G

8 G. Cesàro continuity

A sequence $\{x_n\}$ of real numbers is said to be Cesàro convergent to x_0 if the sequence of its averages has the limit x_0; explicitly if

$$\lim_{n\to\infty} \frac{x_1 + x_2 + \cdots + x_n}{n} = x_0.$$

This is an important and justly famous concept. The first theorem of the subject is that ordinary convergence implies Cesàro convergence. That is: if $\lim_{n\to\infty} x_n = x_0$, then the sequence $\{x_n\}$ is Cesàro convergent to x_0. (The idea of proof is this: if x_n is within ε of x_0 as soon as $n > n_0$, then write the average

$$\frac{x_1 + x_2 + \cdots + x_n}{n}$$

as a sum of two terms, the first involving the first n_0 of the x's, and the second the others; the first summand, with a fixed numerator and n in the denominator, becomes arbitrarily small when n becomes large, and the second summand gets within ε of x_0.)

The importance of the concept comes from the fact that the converse is not true: there are many sequences that are not convergent but are Cesàro convergent. Typical examples: $\{0, 1, 0, 1, 0, 1, \ldots\}$ is Cesàro convergent to $\frac{1}{2}$, and $\{1, 2, -3, 1, 2, -3, 1, 2, -3, \ldots\}$ is Cesàro convergent to 0. A different kind of example is obtained by putting $x_n = 0$ except when n is a power of 10, and $x_n = k$ when $n = 10^k$; in that case the sequence is unbounded but a little mild attention to arithmetic shows that it is Cesàro convergent to 0.

What is an example of a sequence that is not Cesàro convergent? An easy one is given by $x_n = n$, but that may be unfair—it's too unbounded. What is an example of a bounded sequence that is not Cesàro convergent? One way to get one is to alternate longer and longer sequences of 0's and 1's: start with one 0, say, then put several 1's, then many 0's, then very many 1's, and so on. Care with the arithmetic can make the words "several, many, very many, etc." precise in such a way that the sequence of averages has every number in the interval $[0, 1]$ as a limit point.

It can be useful to have a symbol to denote Cesàro convergence; one possibility is to write

$$x_n \to x_0 \ (C)$$

to indicate that $\{x_n\}$ is Cesàro convergent to x_0. A related concept is that of Cesàro continuity. A function f is called Cesàro continuous at a point x_0 in case $x_n \to x_0 \ (C)$ implies that $f(x_n) \to f(x_0) \ (C)$. Trivial example: if $f(x) = x$ for all x, then f is Cesàro continuous at every point. Almost as easy: if $f(x) = Ax + B$, then f is Cesàro continuous at every point.

Problem 8 G. *At which points are the functions defined by* $f(x) = x^2$ *and* $g(x) = \sqrt{x}$ *Cesàro continuous?*

Comment. The relation between ordinary continuity and Cesàro continuity is not obvious. Is a continuous function Cesàro continuous? Is a Cesàro continuous function continuous? The questions are intended to be asked with a universal quantifier: is *every* continuous function Cesàro continuous, and is *every* Cesàro continuous function continuous? The questions can be asked about continuity at one point, or about unqualified continuity, meaning simultaneous continuity at every point.

The logic of the two concepts is not trivial. More sequences are Cesàro convergent than are convergent. The effect of that observation on continuity is that a Cesàro continuous function must "do something" at more sequences than a just plain continuous function, but what it must do is less restrictive. The hypotheses and the conclusions involved in the definitions of the two kinds of continuity pull against one another —passing from ordinary continuity to Cesàro continuity both the hypothesis and the conclusion

$$x_n \to x_0 \ (C) \quad \text{implies that} \quad f(x_n) \to f(x_0) \ (C)$$

are weakened. It is difficult to predict who will win.

8 H. Cantor function

8 H

The Cantor function is not as famous as the Cantor set, but it's at least as spectacular—it has weird properties that become less and less weird as we get used to them.

Let's begin with a reminder. The Cantor set is best defined by defining its complement. Consider the unit interval $[0, 1]$ and in it the set that consists of the open middle third

$$\left(\frac{1}{3}, \frac{2}{3}\right)$$

together with the open middle thirds of the two closed intervals on either side of it, that is

$$\left(\frac{1}{9}, \frac{2}{9}\right) \cup \left(\frac{7}{9}, \frac{8}{9}\right),$$

together with the open middle thirds of the four closed intervals that remain, that is

$$\left(\frac{1}{27}, \frac{2}{27}\right) \cup \left(\frac{7}{27}, \frac{8}{,27}\right) \cup \left(\frac{19}{27}, \frac{20}{27}\right) \cup \left(\frac{25}{27}, \frac{26}{27}\right),$$

and so on forever. The complement of all that is the Cantor set. Since (easy exercise) the sum of the lengths of all the open middle thirds that make up the complement of the Cantor set is exactly 1, that is, the complement of the Cantor set is a set of measure 1 in $[0, 1]$, it follows that the Cantor set has measure 0. It is, nevertheless, a large set in the sense of cardinal numbers, meaning that it is uncountable.

The Cantor function, let us call it φ, is defined on the entire interval $[0, 1]$, usually (and here too) by defining it first on the complement of the Cantor set and then waving one's hand about the rest. Specifically:

$$\varphi(x) = \frac{1}{2} \quad \text{when } x \in \left(\frac{1}{3}, \frac{2}{3}\right),$$

$$\varphi(x) = \frac{1}{4} \quad \text{when } x \in \left(\frac{1}{9}, \frac{2}{9}\right) \quad \text{and}$$

$$\varphi(x) = \frac{3}{4} \quad \text{when } x \in \left(\frac{7}{9}, \frac{8}{9}\right),$$

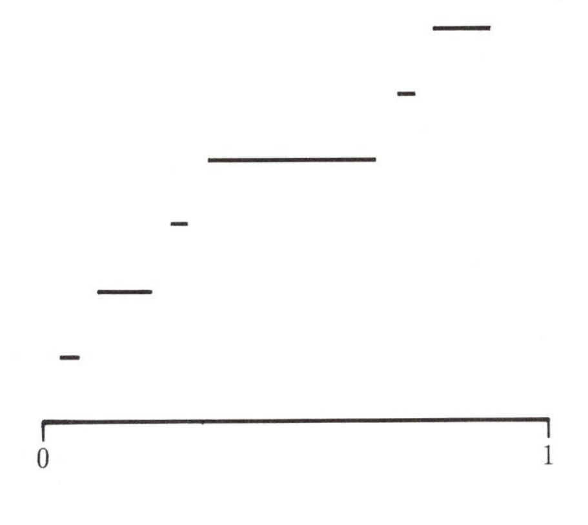

FIGURE 22

and so on. Is it clear what "and so on" means? It means, for instance, that on the four open middle thirds that come next φ is defined to have the values $\frac{1}{8}$, $\frac{3}{8}$, $\frac{5}{8}$, and $\frac{7}{8}$, and the next time, on the next eight open middle thirds, φ will take the eight values with 16 in the denominator and the odd positive integers below 16 in the numerator. The inductive repetition of this process is clear enough and it defines the Cantor function on the complement of the Cantor set. To define the function on the Cantor set itself, the thing to say is that there is a way of doing that so that the resulting function is continuous, and, what's more, there is only one such way—consider it said.

What's the integral of the Cantor function? In other words, what's the area under the curve? The quick way to answer is to say that there is up-down symmetry: the area over the curve (but, of course, below the horizontal line $y = 1$) is equal to the area under the curve. Conclusion: they are both equal to $\frac{1}{2}$. A slower way to answer, but a way that has some value, is to sum the pertinent infinite series:

$$\frac{1}{3}\cdot\frac{1}{2}+\frac{1}{9}\left(\frac{1}{4}+\frac{3}{4}\right)+\frac{1}{27}\left(\frac{1}{8}+\frac{3}{8}+\frac{5}{8}+\frac{7}{8}\right)+\cdots$$
$$=\sum_{n=1}^{\infty}\frac{1}{3^n}\left(\frac{1}{2^n}+\frac{3}{2^n}+\cdots+\frac{2^n-1}{2^n}\right),$$

and come to the same conclusion.

The Cantor set (and, correspondingly, the Cantor function) can and should be generalized—the generalized versions can often serve as examples and counterexamples. The original Cantor set has measure 0; the generalized ones have positive measure. Suppose, to be specific, that $\{\mu_n\}$ is a sequence of numbers such that $0 < \mu_n < 1$, and such that

$$\sum_{n=1}^{\infty} 2^{n-1}\mu_n = 1 - \alpha, \quad \text{where } 0 \leqq \alpha < 1.$$

Use the μ's the same way as the powers of $\frac{1}{3}$ are used in the definition of the Cantor set. That is: form the open middle interval of length μ_1 in $[0, 1]$, then add on the two open middle intervals of length μ_2 that are in the two closed intervals on either side of it, continue with the open middle intervals of length μ_3 of the four closed intervals left, and so on forever. The complement of all that is the generalized Cantor set. Since the complement has measure $1 - \alpha$, the generalized Cantor set has measure α. For the original set $\mu_n = \frac{1}{3^n}$ and, therefore, $\alpha = 0$.

The Cantor function can be defined exactly the same way now as before: let its values be the successive odd multiples of the successive powers of $\frac{1}{2^n}$ exactly as before.

Problem 8 H. *What is the length of the graph of the (generalized) Cantor function?*

Comment. Emphasis: what's being asked for is not the area under the curve, but its length. The area is easily calculated by the summation method mentioned for the case $\mu_n = \frac{1}{3^n}$; the answer turns out to be $\frac{1}{2}(1 - \alpha)$. The area between the curve and the horizontal line $y = 1$ is $\frac{1}{2}(1 + \alpha)$.

8 I

8 I. Fixed points of entire functions

The subject called analysis is the biggest part of mathematics—it embraces elementary calculus, powerful applications, abstract operator theory, and, in particular, the theory of functions of a complex variable. The latter is a beautiful subject that deserves a lot of study—but all it can get here is a brief courteous nod by way of a couple of small problems. Hope: even non-specialists will find these problems appealing.

The most natural generalization of polynomials is to "long" polynomials, that is to power series $\sum_{n=0}^{\infty} a_n z^n$. Convergence questions arise, and the nicest way to get around them, at least at the beginning, is to assume that the series are as well behaved as they can be, that is, that they converge for every complex value of z. If that is the case, then their sums $f(z)$ are important special kinds of analytic functions, the kind called entire functions.

Does an equation of the form $p(z) = 0$, where p is a polynomial, always have a solution? Most of us are conditioned to say yes, but that's not right—the answer is no for polynomials such as the one defined by $p(z) = 17$. The fundamental theorem of algebra tells us that the answer is yes for all others; it can be no only for constants different from zero. Does an equation of the form $f(z) = 0$, where f is an entire function (not a constant), always have a solution? No—the standard example is the exponential function defined by $f(z) = e^z$.

Does a non-constant polynomial mapping always have a fixed point? That is, if p is a polynomial, not a constant, must the equation $p(z) = z$ have a solution? Here is another chance to make a small slip—it is tempting just to form the difference $p(z) - z$, recognize that it is a polynomial, and then solve the equation $p(z) - z = 0$. The reason that's a slip is that $p(z) - z$ could be a non-zero constant—in other words $p(z)$ could be a translation such as $z + 17$. The correct answer is that every polynomial mapping that is not a translation has a fixed point. Does mapping by an entire function that is not a translation always have a fixed point? No—the fixed point question for entire functions reduces to the zero question for entire functions, just as it does for polynomials. Specifically: if $f(z) = z + e^z$, then f is an entire function, and if it had a fixed point, then e^z would have a zero.

If p is a polynomial, not a translation, must the equation $p(p(z)) = z$ always have a solution? Sure, just as before: $p(p(z)) - z$ is a polynomial and must have a zero. The assumption that $p(z)$ is not a translation implies that $p(p(z)) - z$ cannot be a non-zero constant (proof?) and all is well. What happens to the entire function analogue of this question? The simple trick of looking at e^z or $z + e^z$ doesn't seem to give any immediate information—some other idea is needed.

Problem 8 I. *If f is an entire function that is not a translation, must the equation $f(f(z)) = z$ always have a solution?*

8 J

8 J. Monic polynomials

In many respects the behavior of a polynomial is like the behavior of its leading term (which is the term of highest degree that actually occurs). If for instance the leading coefficient a_n of a polynomial $p(z) = \sum_{j=0}^{n} a_j z^j$ has large absolute value (in this situation that can be interpreted to mean that $|a_n| > 1$), then it seems likely (doesn't it?) that the mapping p will send some points of the closed unit disk out of that disk (will increase their absolute values). If, on the other hand, the leading coefficient is small ($|a_n| < 1$), then there is a chance that p will map the closed unit disk into itself. The problematic case is that of monic polynomials—the ones whose leading coefficient is equal to 1. (There is no real loss in assuming that a_n itself is 1, not just the absolute value $|a_n|$—multiplying the polynomial by a constant of absolute value 1 doesn't alter its "size".) The easiest monic polynomials to study are the powers, $p_n(z) = z^n$; it is clear that they map the disk into itself. How typical are they?

Problem 8 J. *Which monic polynomials map the unit disk into itself?*

8 K

8 K. Semiuniversal chords

For a number c in $[0, 1]$ and a continuous function f on $[0, 1]$, with $f(0) = f(1) = 0$, it can happen that c is a chord of f but $1 - c$ is not. To see a non-trivial example, look at the function constructed in Solution 8 F to exclude the chord $c = \frac{2}{5}$ and observe that $\frac{3}{5}$ is a chord of it. Interchange the roles of $\frac{2}{5}$ and $\frac{3}{5}$ to become convinced that it can also happen the other way. For all we know so far it might also happen that neither c nor $1 - c$ is a chord of f, and, for other c's and f's, that both c and $1 - c$ are chords of f. The latter is what happens, for instance, in the trivial case $f(x) \equiv 0$, or, in the less trivial case $f(x) = x - x^2$. How likely is the former?

If all that is known about a number c and a function f is that either c or $1 - c$ is a chord of f, so that, roughly speaking, the available information is half the amount needed to come to a firm conclusion, it might be reasonable to call c a semichord of f. To continue the terminology along the same lines, let us agree to say that a number c is a semiuniversal chord (universal semichord?) if either c or $1-c$ is a chord of every f (so that, in particular, every universal chord is a semiuniversal chord).

Problem 8 K. *Which numbers are semiuniversal chords?*

8 L. Bounded polynomials

8 L

Problem 8 L. *If a real polynomial is bounded from below, does it necessarily attain its greatest lower bound?*

Comment. A "real" polynomial is one with real coefficients. Such a thing obviously induces (is?) a real-valued function, and the question is one about the analytic properties of that function.

8 M. Polynomial limits

8 M

Problem 8 M. *If a real-valued function on the real line is the limit of a uniformly convergent sequence of polynomials, does it follow that it is differentiable at every point?*

8 N. Harmonic series, primes

8 N

The word "series" is sometimes left undefined in calculus courses, and, therefore, it sometimes confuses students. It is an easy word to define: it is just a synonym for a particular kind of sequence, a synonym that intends to put emphasis on the way the sequence is formed. The infinite series denoted by $\sum_{n=1}^{\infty} a_n$ is, by definition, the infinite sequence

$$a_1, a_1 + a_2, a_1 + a_2 + a_3, \ldots,$$

that is, the sequence whose kth term is

$$\sum_{n=1}^{k} a_n$$

and the notation is nothing more than an efficient reminder of how the terms of the sequence are to be constructed (starting from a given sequence $a_1, a_2, a_3, \ldots$).

One of the most famous infinite series is

$$\sum_{n=1}^{\infty} \frac{1}{n},$$

the so-called harmonic series. An exercise about it that usually appears early in textbook treatments is that its partial sums behave logarithmically. The precise statement is that the sequence

$$1 + \frac{1}{2} + \cdots + \frac{1}{k} - \log k$$

tends to a limit as k tends to ∞. The proof involves nothing more profound than comparing the sum with its "continuous" version

$$\int_1^k \frac{dx}{x}$$

and a slight juggling with inequalities. (The limit is called Euler's constant; its value is .577... .) It follows immediately that the harmonic series diverges:

$$\sum_{n=1}^{\infty} \frac{1}{n} = \infty.$$

What's the intuitive content of that divergence statement? Presumably it is that if we add "many" terms of the harmonic series, we get a "large" sum? How many? How large? A little numerical experimentation (using nothing more profound than the elementary inequalities mentioned above) shows that if we had a computer with infinite precision, and started calculating the partial sums of the harmonic series at the rate of one term per second, it would take us something over 28 million years to reach the sum 35. At the rate of a thousand terms per second that comes to only twenty eight thousand years, and even at the improbable rate of a million terms per second, it would take twenty eight years. What conclusion is the calculation of those partial sums likely to suggest: convergence or divergence?

Granted, however, that the harmonic series diverges, there is some profit to be gained by looking at some of its subseries. A sufficiently "small" subseries will, of course, converge. Is the subseries given by the primes small enough?

Problem 8 N. *If P is the set of all primes, is the sum*

$$\sum_{n \in P} \frac{1}{n}$$

finite or infinite?

Comment. A symbol such as $\sum_{n \in P} \frac{1}{n}$ indicates the sum of the reciprocals of the integers in the indicated set P. Thus, for instance, if N denotes the set of all positive integers, then the statement that the harmonic series diverges can be expressed by writing $\sum_{n \in N} \frac{1}{n} = \infty$.

8 O. Rational exponentials

8 O

The number $e = \sum_{n=0}^{\infty} \frac{1}{n!}$ is known to be transcendental, and hence, in particular, irrational; how catching is that disease?

Problem 8 O. *For each sequence $\{\varepsilon_n\}$ of plus and minus 1's, write*

$$e(\{\varepsilon_n\}) = \sum_{n=0}^{\infty} \frac{\varepsilon_n}{n!}.$$

For which sequences $\{\varepsilon_n\}$ are the numbers $e(\{\varepsilon_n\})$ rational?

8 P. Harmonic series, values

8 P

Problem 8 P. *Does there exist a sequence $\{\varepsilon_n\}$ of plus and minus 1's such that the infinite series*

$$\sum_{n=1}^{\infty} \frac{\varepsilon_n}{n}$$

converges and has the value e?

8 Q. Harmonic series, gaps

8 Q

Most numbers are composite, one would think, and, therefore, it would seem that very few numbers are primes—that's what makes it surprising that the omission from the harmonic series of the reciprocals of the composite numbers leaves it divergent. In the decimal representation of some numbers the digit 0 occurs and in others it does not: which are there more of? The reciprocals of one or the other of the two kinds (or both) must constitute a divergent series—which?

Problem 8 Q. *If A is the set of those positive integers in whose decimal representation* 0 *does not occur, is the series $\sum_{n \in A} \frac{1}{n}$ convergent or divergent?*

CHAPTER 9

MATRICES

9 A. Unions of subspaces

9 A

We usually think of linear algebra as the part of mathematics that studies vector spaces, but we shouldn't—according to the right description, linear algebra is the part of mathematics that studies linear transformations. Still and all, vector spaces is where all the action takes place, and it is not inappropriate to begin a section of problems in linear algebra with some attention to spaces.

The union of two lines in the plane is, typically, a cross, something like an $\times$—or, in an extreme case, when the two given lines coincide, it is just a line. Similar comments can be made about the union of two planes in space—geometric intuition is still working. Geometric intuition is likely to stop working when it is consulted about the union of two subspaces of a 19-dimensional space—say a 17-dimensional one and an 18-dimensional one.

Problem 9 A. *Can a real vector space be the union of a finite number of proper subspaces?*

Comment. A "real" vector space is one for which the field of scalars is the field of real numbers. "Proper" means not equal to the whole space.

9 B

9 B. Simultaneous complements

The complement of a subset S of a set V is the set T of all elements of V that are not in S. Equivalently: the complement T of S is characterized by the properties

$$S \cap T = \varnothing$$

(that is, S and T are disjoint), and

$$S \cup T = V,$$

(that is, S and T add up to V). A similar definition can be made for subspaces of a vector space $\mathbb{V}$—similar, but not identical. Each of two subspaces $\mathbb{S}$ and $\mathbb{T}$ of $\mathbb{V}$ is called a complement of the other in case $\mathbb{S}$ and $\mathbb{T}$ are "disjoint" and $\mathbb{S}$ and $\mathbb{T}$ "add up" to $\mathbb{V}$—but "disjoint" and "add up" have different meanings from the ordinary set-theoretic ones.

The empty set $\varnothing$ is characterized by the property that it is a subset of every set. Which subspace of $\mathbb{V}$ is a subspace of every subspace? Answer: the subspace $\mathbb{O}$ that consists of the zero vector only. In view of that observation, some authors say that two subspaces are disjoint if their intersection is $\mathbb{O}$.

The union of two sets S and T is characterized by the property that it is the smallest set that both S and T are subsets of. If $\mathbb{S}$ and $\mathbb{T}$ are subspaces of $\mathbb{V}$, what is the smallest subspace of $\mathbb{V}$ that both $\mathbb{S}$ and $\mathbb{T}$ are subspaces of? The answer is usually called the span of $\mathbb{S}$ and $\mathbb{T}$. It is a familiar fact, easy to prove, that if $\mathbb{V}$ is finite-dimensional, and if $\mathbb{S}$ and $\mathbb{T}$ are subspaces of $\mathbb{V}$, then the span of $\mathbb{S}$ and $\mathbb{T}$ consists of the set of all vectors of the form $s+t$, with s in $\mathbb{S}$ and t in $\mathbb{T}$. (The set of all such sums $s+t$ is usually called the sum of the two subspaces, and is denoted by $\mathbb{S}+\mathbb{T}$. The fact just stated is that, in finite-dimensional spaces, the span of two subspaces is equal to their sum. Caution: in vector spaces that are not finite-dimensional that equality is not necessarily true.)

In view of the preceding two paragraphs, two subspaces $\mathbb{S}$ and $\mathbb{T}$ of a vector space are called complements of one another if they are disjoint, that is

$$\mathbb{S} \cap \mathbb{T} = \mathbb{O},$$

and if their span is $\mathbb{V}$; in finite-dimensional spaces the latter condition can be written in the form

$$\mathbb{S} + \mathbb{T} = \mathbb{V}.$$

A subset S of a set V has only one complement in V; a subspace $\mathbb{S}$ of a vector space $\mathbb{V}$ might have many complements. The x-axis in the plane is a subspace of the plane, and the y-axis is a complement of it, but so also is the line $\{(x, y) : x = y\}$ (the 45° diagonal line), as well as the line $\{(x, y) : x + y = 0\}$, and, in fact, as well every other line through the origin distinct from the x-axis itself. More is true: any two distinct lines through the origin, regarded as subspaces of the plane $\mathbb{R}^2$, are complements. Similarly, if $\mathbb{S}$ is a plane through the origin in $\mathbb{R}^3$ and $\mathbb{T}$ is a line through the origin and not in the plane $\mathbb{S}$, then $\mathbb{S}$ and $\mathbb{T}$ are complements. In other words, it is easily possible for distinct subspaces to have common (simultaneous) complements.

Just how easy is it? Is it possible in a 19-dimensional space for a 17-dimensional and an 18-dimensional subspace to have a simultaneous complement? No—a moment's thought about the meaning of dimension will reveal that if $\mathbb{S}$ and $\mathbb{T}$ are complementary subspaces of an n-dimensional vector space $\mathbb{V}$, then the dimensions of $\mathbb{S}$ and $\mathbb{T}$ add up to n. Every complement of a 17-dimensional subspace of a 19-dimensional vector space must have dimension 2, and a 2-dimensional subspace cannot be a complement of an 18-dimensional one. What else can go wrong?

Problem 9 B. *If a finite number of subspaces of a finite-dimensional real vector space all have the same dimension, must they necessarily have a simultaneous complement?*

9 C. Square roots of matrices

9 C

One reason why matrices (linear transformations) are fun is that they behave like numbers—but not quite. Here is an example: every complex number has a (complex) square root; does every (complex) matrix have a square root? Is the meaning of the words clear? If A and B are matrices such that $A = B^2$, then, naturally, B is called a square root of A.

The answer is no, not every matrix has a square root; an easy example is $A = \left(\begin{smallmatrix} 0 & 1 \\ 0 & 0 \end{smallmatrix}\right)$. Why doesn't A have a square root? Presumably one way to establish that negative statement would be to consider an "unknown" 2×2 matrix B (with four unknown entries), square it, get four equations in four unknowns, and prove that those equations have no solution. That's not the recommended intelligent procedure; thinking is better than computing. One way to think is to realize that $A^2 = 0$ (the zero matrix), and that, therefore, if A had a square root B, then it would be the case that $B^4 = 0$ (but, of course, $B^2 \neq 0$, because $A \neq 0$). Can that happen? Can there be a 2×2 matrix whose 4th power is 0 but whose 2nd power is not?

The question makes contact with a sophisticated and important part of linear algebra, namely the theory of polynomials that annihilate a matrix. To say that a polynomial $p(\lambda)$ annihilates A means that the result of substituting A for the variable in p is the zero matrix: that is, $p(A) = 0$. The most famous assertion along these lines is called the Hamilton–Cayley equation: it asserts that if $p(\lambda) = \det(A - \lambda)$ (the characteristic polynomial of A), then $p(A) = 0$. (In a context such as $A - \lambda$ the symbol "λ" denotes λ times the identity transformation.) If A is an $n \times n$ matrix, then the degree of the characteristic polynomial of A is n. It can happen that there exist polynomials of degree lower than n that also annihilate A; the monic polynomial of lowest degree that does so is called the minimal polynomial of A. Typical example: if

$$A = \begin{pmatrix} 0 & 0 & 0 & 0 \\ 0 & 0 & 0 & 0 \\ 0 & 0 & 1 & 0 \\ 0 & 0 & 0 & 1 \end{pmatrix},$$

then the characteristic polynomial of A is $\lambda^2(1 - \lambda)^2$ and the minimal polynomial of A is $\lambda(1 - \lambda)$. The minimal polynomial is always a factor of the characteristic polynomial, and, in fact, it is a factor of every polynomial that annihilates A. Trivial but useful consequence: the degree of the minimal polynomial of an $n \times n$ matrix is never more than n.

Once these things about polynomials are recalled, the facts about the square root example become transparent. Since the polynomial $p(\lambda) = \lambda^4$ annihilates B, it follows that the minimal polynomial of B must be a factor of λ^4, and therefore the minimal polynomial of B must be a power of λ. Since B is not the zero matrix (because its square is A), the minimal polynomial is not the first power of λ, and therefore since B

is a 2×2 matrix, its minimal polynomial must be λ^2. But that says that $B^2 = 0$, which contradicts the assumption that B is a square root of A.

Very well then, matrices might fail to have square roots, but some experience with ones that do and ones that don't can only be useful.

Problem 9 C. *Does either of the matrices*

$$\begin{pmatrix} 0 & 0 & 1 \\ 0 & 0 & 0 \\ 0 & 0 & 0 \end{pmatrix} \quad \textit{and} \quad \begin{pmatrix} 0 & 1 & 0 \\ 0 & 0 & 0 \\ 0 & 0 & 0 \end{pmatrix}$$

have a square root?

9 D. Inverses and polynomials

9 D

Given a matrix A and a function f, it sometimes makes sense to consider $f(A)$. The pleasantest and simplest instances of this phenomenon are polynomials in a matrix; they are easy to work with. Other functions of matrices can occur however (such as inverses, transposes, adjoints) and it is sometimes useful to know whether or not they can be expressed in terms of polynomials. Here is a sample.

Problem 9 D. *If A is an invertible linear transformation on a finite-dimensional vector space, does there necessarily exist a polynomial p such that $A^{-1} = p(A)$?*

9 E. Extra eigenvectors

9 E

It is true enough that in a 2-dimensional vector space no set of three or more vectors can be linearly independent, but could there exist three vectors such that *every two* of them are linearly independent? Sure, that's not hard; one example in $\mathbb{C}^2$ is the set consisting of the three vectors $(1, 0)$, $(0, 1)$, and $(1, 1)$. Could such vectors all be eigenvectors of one and the same linear transformation A? Yes—just consider $A = 1$, or more generally let A be any scalar multiple of the identity. That deserves to be called a trivial example; what are the non-trivial ones?

Problem 9 E. *What are all possible examples of linear transformations on an n-dimensional space that have $n+1$ eigenvalues such that every subset of n of them is linearly independent?*

9 F

9 F. Even matrices

Problem 9 F. *If a 7×7 matrix A has all its diagonal entries equal to 0 and all its non-diagonal entries equal to either $+1$ or -1, can its determinant be 0? What about an 8×8 matrix—under what additional conditions can A be invertible?*

9 G

9 G. Null power matrices

Eigenvalues tend to reflect the properties of the matrices they belong to—here is an instance. If a real or complex square matrix A is such that $A^n \to 0$ as $n \to \infty$, then every eigenvalue of A is strictly less than 1 in absolute value. Reason: if $Ax = \lambda x$, then $A^n x = \lambda^n x$, and therefore $\lambda^n \to 0$. How good is the correspondence between the properties of the matrix and the properties of its eigenvalues? Specifically: is the converse of the statement just made also true?

Problem 9 G. *If $|\lambda| < 1$ for every eigenvalue λ of a matrix A, does it follow that $A^n \to 0$ as $n \to \infty$?*

9 H

9 H. Moving averages

The first two terms of the Fibonacci sequence are 0 and 1; what would happen if those initial values were replaced by other real numbers? In the Fibonacci sequence each term after the first two is the sum of its two predecessors; what would happen if "sum" were replaced by "average"? Can the sequences obtained by such modifications of Fibonacci tend to infinity? Can they be bounded? Can they converge? And what happens if the Fibonacci sequence is generalized even more by starting with an arbitrary sequence of length k, say (instead of length 2)?

Problem 9 H. *If $\{a_0, a_1, \ldots, a_{k-1}\}$ is a sequence of k numbers (where k is an arbitrary but from now on fixed positive integer),* *and if* a_{n+k} *is defined, for* $n = 0, 1, 2, \ldots$ *by*

$$a_{n+k} = \frac{1}{k}\sum_{j=0}^{k-1} a_{n+j},$$

under what conditions on $\{a_0, a_1, \ldots, a_{k-1}\}$ *is the sequence* $\{a_{n+k}\}$ *convergent?*

9 I. Square roots and polynomials

9 I

The inverse of an invertible linear transformation on a finite-dimensional vector space is always a polynomial in the given transformation (see 9 D); what about the adjoint? (At this point we abandon "pure" linear algebra and turn to a few problems involving inner product spaces and transformations defined on them—from now on words like "adjoint", and "Hermitian", and "normal" can be used. The present question could have been asked as a pure matrix question even before this change of orientation: it makes sense to ask whether the transpose, or the conjugate transpose, of a matrix is or is not a polynomial in it.)

The answer to the adjoint question is different from the answer to the inverse question: the adjoint of a transformation may fail to be a polynomial in it. There are easy counterexamples;

$$A = \begin{pmatrix} 0 & 1 \\ 0 & 0 \end{pmatrix}$$

is one of them. Indeed: every polynomial in A is of the form

$$\begin{pmatrix} \alpha & \beta \\ 0 & \alpha \end{pmatrix},$$

whereas

$$A^* = \begin{pmatrix} 0 & 0 \\ 1 & 0 \end{pmatrix}$$

is not of that form.

What about square roots in place of inverses or adjoints? Not every matrix has a square root, so a little care has to be exercised in asking the question. The safest way to ask it is to restrict attention to positive (= positive definite, or, in classically correct pedantic language, non-negative semi-definite) transformations; general theory guarantees the existence of a unique positive square root for each of them, and it makes sense to ask whether for, such transformations A, the square root $\sqrt{A}$ is or is not a polynomial in A. The answer is almost obviously yes—positive transformations are, in particular, Hermitian, and can, therefore, be "diagonalized". That is: there exists a basis with respect to which the matrix of A is diagonal. The square root of a (positive) diagonal matrix is the diagonal matrix whose diagonal entries are the (positive) square roots of the given one. Since polynomials can be found that take arbitrarily prescribed values at arbitrarily prescribed finite sets, the square root of a positive matrix is a polynomial in it. That's what the general theory says, but are such comments helpful in actually finding and exhibiting square roots?

Problem 9 I. *What is the positive square root of*

$$\begin{pmatrix} 2 & 1 \\ 1 & 1 \end{pmatrix}?$$

Comment. The answer can be found by computing, but, as usual, it is advisable to look for more intelligent ways of getting it. What is really wanted is a formula for $\sqrt{A}$ in terms of A, a polynomial formula that works for every positive linear transformation A on $\mathbb{C}^2$, and then an application of that formula to the particular matrix at hand.

9 J

9 J. Linear isometries

The action of

$$A = \frac{1}{\sqrt{2}} \begin{pmatrix} 1 & 1 \\ 1 & -1 \end{pmatrix}$$

on $\mathbb{R}^2$ is linear, of course, and it has also a considerably more rare geometric property, namely that it is isometric. Meaning: the distance betweeen Ax and Ay is always the same as the distance between x and y. More generally, orthogonal transformations (on real inner product spaces) are isometries.

Is every isometry linear? That is: if $\mathbb{V}$ is a finite-dimensional real inner product space and if U is a mapping of $\mathbb{V}$ into itself such that

$$\|Ux - Uy\| = \|x - y\|$$

for all x and y, does it follow that U is a linear transformation? The answer is obviously no: just start with a linear isometry and ruin it. That is: given a linear isometry V, form $U = V + 1$; the resulting U is just as isometric as V, but it is no longer linear. It can be argued, however, that the way that U is different from a linear transformation is not very deep; is it possible to construct a truly non-linear isometry?

Problem 9 J. *If U is an isometry (on a finite-dimensional inner product space) that maps 0 onto 0, must U be linear?*

9 K. Perturbation of the identity

Consider on $\mathbb{R}^3$ the linear transformation B defined by

$$B(x_1, x_2, x_3) = (x_1, 0, 0);$$

the matrix of B with respect to the usual basis

$$\{(1,0,0),(0,1,0),(0,0,1)\}$$

is

$$\begin{pmatrix} 1 & 0 & 0 \\ 0 & 0 & 0 \\ 0 & 0 & 0 \end{pmatrix}.$$

Alternatively, consider on $\mathbb{R}^2$ the linear transformation C defined by

$$C(x_1, x_2) = (x_1 + x_2, -x_1 - x_2),$$

with matrix

$$\begin{pmatrix} 1 & 1 \\ -1 & -1 \end{pmatrix}.$$

Both of these examples are instances of a very special way of manufacturing linear transformations; that way can be described as follows. Choose any two vectors u and v in a finite-dimensional inner product space, and define a linear transformation A (sometimes denoted by $u \otimes v$) by the equation

$$Ax = (x, v)u.$$

In other words, the linear transformation A sends every vector x in the space onto a scalar multiple of the fixed vector u; the scalar is determined by the fixed vector v (and, of course, by the variable x). For the example B, the vectors can be chosen as $u = v = (1, 0, 0)$, and for C they can be chosen as $u = (1, -1)$ and $v = (1, 1)$.

The very definition of rank implies that the linear transformation A has rank 1 no matter what u and v are. Transformations of rank 1 are "small" in some sense, and the result of changing a given transformation by a small amount (in varying senses of small) is sometimes called a perturbation of it. How much change can a perturbation by a transformation of rank 1 affect? Consider, as an example, the identity transformation (on a finite-dimensional inner product space). Its determinant is 1; how much can its determinant change under a perturbation of rank 1?

> **Problem 9 K.** *Under what conditions on the vectors u and v can the transformation $1+A$, where A is defined by $Ax = (x, v)u$, have negative determinant?*

Comment. Just like the question about adjoints in the preliminary discussion of 9 I, this question could have been asked in the language of pure matrix theory, but, in both cases, the language of inner product spaces puts the question into a more comfortable context.

9 L

9 L. Commuting normal transformations

Normal transformations (on inner product spaces) are the most useful common generalization of the important classes of Hermitian and unitary transformations. (Reminder: A is normal if it commutes with its adjoint, that is $A^*A = AA^*$. Strongest theorem about normal transformations: they are exactly the ones that are unitarily diagonalizable. This is sometimes referred to as the spectral theorem.) The bad news is that a product of normal transformations can fail to be normal. Here is an easy example of two normal transformations (matrices) whose product is not normal no matter in which order the multiplication is carried out:

$$\begin{pmatrix} 1 & 0 \\ 0 & 0 \end{pmatrix} \cdot \begin{pmatrix} 0 & 1 \\ 1 & 0 \end{pmatrix} = \begin{pmatrix} 0 & 1 \\ 0 & 0 \end{pmatrix}$$

and

$$\begin{pmatrix} 0 & 1 \\ 1 & 0 \end{pmatrix} \cdot \begin{pmatrix} 1 & 0 \\ 0 & 0 \end{pmatrix} = \begin{pmatrix} 0 & 0 \\ 1 & 0 \end{pmatrix}.$$

What if they commute?

Problem 9 L. *Is the product of two commutative normal transformations always normal?*

9 M. Norms of normal products

9 M

In general the norm of the product of two transformations depends on the order of the factors. Easy example:

$$\text{if } A = \begin{pmatrix} 1 & 0 \\ 0 & 0 \end{pmatrix} \quad \text{and} \quad B = \begin{pmatrix} 0 & 0 \\ 1 & 0 \end{pmatrix},$$

then

$$AB = \begin{pmatrix} 0 & 0 \\ 0 & 0 \end{pmatrix} \quad \text{and} \quad BA = \begin{pmatrix} 0 & 0 \\ 1 & 0 \end{pmatrix},$$

so that

$$\|AB\| = 0 \quad \text{and} \quad \|BA\| = 1.$$

In this example A is normal, but B is not. Can this "misbehavior" occur when both A and B are normal?

Problem 9 M. *Can the equality $\|AB\| = \|BA\|$ fail for normal transformations A and B?*

9 N. Law of exponents

9 N

Matrices behave like numbers, yes, but not quite—that has been said before. One place where the difference shows up startlingly is in the properties of the exponential function. For any matrix A, it makes sense to speak of e^A; the most obvious definition of e^A is via power series. The first question that is likely to arise after a proof that e^A makes sense is whether it satisfies the normal law of exponents—in other words whether or not it is true for all A and B that

$$e^{A+B} = e^A e^B.$$

If A and B commute, the answer is yes, and the proof in that case is essentially the same as it is for numerical exponents; all it takes is some patient juggling with the infinite series definition. In case A and B do not commute, that proof breaks down; does the conclusion break down?

Problem 9 N. *Do there exist matrices A and B such that*

$$e^{A+B} \neq e^A e^B?$$

9 O

9 O. Necessary commutativity

Commutativity is a sufficient condition for the law of exponents to hold and without commutativity it may fail to hold. Is commutativity in fact a necessary condition for the law of exponents?

Problem 9 O. *If $e^{A+B} = e^A e^B$, does it follow that A and B commute?*

9 P

9 P. Hermitian law of exponents

If A and B commute, then e^{A+B} must be equal to $e^A e^B$, but if A and B do not commute, then e^{A+B} may or may not be equal to $e^A e^B$. What if A and B are "good" linear transformations—is the exponential law equivalent to commutativity in that case? One way to interpret "good" is as Hermitian.

Problem 9 P. *If A and B are Hermitian matrices such that $e^{A+B} = e^A e^B$, does it follow that A and B commute?*

9 Q

9 Q. Exponential inequality

Positive matrices behave like positive numbers, but not quite. If, for instance, $0 \leqq A \leqq B$ (in the sense that both A and $B - A$ are positive matrices), then it does not follow that $A^2 \leqq B^2$. Example:

$$A = \begin{pmatrix} 1 & 0 \\ 0 & 0 \end{pmatrix}, \quad B = \begin{pmatrix} 2 & 1 \\ 1 & 1 \end{pmatrix}.$$

On the other hand, if $0 \leqq A \leqq B$, then it does follow that

$$\sqrt{A} \leqq \sqrt{B};$$

that's a standard exercise in linear algebra.

The most interesting question of this kind about matrix exponentiation concerns the monotoneness of the exponential function. If a and b are positive numbers with $a \leqq b$, then $e^a \leqq e^b$; does that assertion remain true for matrices?

Problem 9 Q. *If A and B are matrices such that* $0 \leqq A \leqq B$, *does it follow that* $e^A \leqq e^B$?

Comment. For which functions f is it true that from $0 \leqq A \leqq B$ it does follow that $0 \leqq f(A) \leqq f(B)$? A characterization of such functions is known, and it is deep and hard. The present question is whether or not the exponential function is one of them.

CHAPTER 10

ALGEBRA

10 A. Reals mod 1

10 A

The set $\mathbb{R}$ of all real numbers is an additive group. The subset $[0,1)$ (a half-open unit interval) is not a group, not a subgroup of $\mathbb{R}$, but it can be made into a group by slightly redefining addition. If x and y are elements of $[0,1)$ such that $x+y<1$, define their new sum $x\,(+)\,y$ to be the same as the old $x+y$; if $x+y \geqq 1$, define $x\,(+)\,y$ to be $x+y-1$. It is easy to verify that the result converts $[0,1)$ into an additive group; it is called the group of real numbers modulo 1 (abbreviated to mod 1). From a more elegant point of view that group can be described as the quotient group $\mathbb{R}/\mathbb{Z}$, where $\mathbb{Z}$ is the subgroup of all integers in $\mathbb{R}$.

Every additive group can be made a ring, in a trivial way, by defining the product of any two elements to be 0 (the additive identity). The set $\mathbb{R}$ is a ring in a non-trivial way because it is a ring with a unity element—or, in other words, the group of all real numbers is the additive group of a ring with unity.

Problem 10 A. *Is the additive group of all real numbers modulo* 1 *the additive group of a ring with unity*?

10 B

10 B. Groups in fields

A ring can be viewed as a group, an abelian group, an additive group, in which a multiplication is defined. A field, in particular, is a ring, a ring in which multiplication is commutative, and a ring, at that, in which multiplication is well behaved. The good behavior is most simply described by saying that, except for the element 0, the elements of a field constitute a multiplicative group. That may be the simplest statement, but it isn't a precise one. The precise statement is that if $\mathbb{F}$ is a field and if $\mathbb{F}^*$ is the set of its non-zero elements, then $\mathbb{F}^*$ is a multiplicative group.

[The symbol $\mathbb{F}^*$ for $\mathbb{F} - \{0\}$—that is, the symbol for the field decorated with an asterisk—is a universally accepted symbol for the multiplicative group of the field. Incidentally: phrases such as "additive group" or "multiplicative group" are just abbreviations: they mean a group in which the operation is called addition and denoted by $+$ or is called multiplication and denoted by $\times$. Correction: the "additive group of a ring" means something specific—namely the ring with multiplication temporarily ignored and addition regarded as the fundamental group operation, and, similarly, the "multiplicative group of a field" means the field with 0 omitted and addition temporarily ignored, with multiplication regarded as the fundamental group operation.]

Some interesting questions can be asked about the groups associated with fields. Consider, for instance, the field $\mathbb{Q}$ of rational numbers, the field $\mathbb{R}$ of real numbers, and the field $\mathbb{C}$ of complex numbers. The middle one of these has a well known and useful property: (1) the additive group of $\mathbb{R}$ is isomorphic to the multiplicative group $\mathbb{R}^{+^*}$ of positive real numbers. (Proof: the function that assigns to each real number x the positive real number e^x is an isomorphism.) Another property is less well known and is sometimes regarded as pathological, namely that (2) the additive group of $\mathbb{R}$ is isomorphic to the additive group of $\mathbb{C}$. (The proof depends on Hamel bases, a mildly delicate piece of set theory.) The first two questions that follow should be contrasted with these two properties of $\mathbb{R}$.

Problem 10 B.

(1) *Is the additive group of $\mathbb{Q}$ isomorphic to the multiplicative group $\mathbb{Q}^{*^+}$ of positive rational numbers?*

(2) *Are the multiplicative groups $\mathbb{R}^*$ and $\mathbb{C}^*$ isomorphic?*

(3) *Is there a field whose additive group is isomorphic to its multiplicative group?*

10 C. Elementary symmetric functions

10 C

If $a_1, a_2, \ldots, a_n$ are elements of a ring, then the elementary symmetric functions formed from them are their sum

$$a_1 + a_2 + \cdots,$$

the sum of their products two at a time

$$a_1a_2 + a_1a_3 + a_2a_3 + \cdots,$$

the sum of their products three at a time

$$a_1a_2a_3 + \cdots,$$

and so on up to their product

$$a_1a_2 \cdots a_n.$$

If the a's are positive real numbers, then their elementary symmetric functions are positive real numbers. Converse?

> **Problem 10 C.** *If all the elementary symmetric functions of a finite set of real numbers are positive, must all the numbers themselves be positive?*

10 D. Polynomial divisibility

10 D

If a polynomial in x vanishes when $x = 2$, then it is divisible by $x - 2$; this is an elementary statement (the remainder theorem, the factor theorem, the division algorithm—the statement can be baptized in several different ways) that is true about polynomials with coefficients in any commutative ring $\mathbb{R}$. Does something like it remain true for polynomials in several variables?

> **Problem 10 D.** *If a polynomial in x and y vanishes when $x = y$, must it be divisible by $x - y$?*

10 E. Non-isomorphic fields

10 E

What's wrong with the following definition: a field is a set with two operations, called addition and multiplication, such that it is an abelian group with respect to addition (with an additive identity called zero), and such that the non-zero elements form an abelian group with respect to multiplication? There is something seriously wrong—the proposed definition has a serious flaw. It is mathematically reprehensible to consider a set endowed with two different structures without demanding a structural connection between them—one is tempted to say that it is immoral to do so.

Consider the patently absurd example of the non-negative integers between 0 and 14 inclusive, with addition defined to be addition modulo 15, and with multiplication defined for the non-zero ones among them as addition (!) modulo 14. Surely no one wants to call that a field. What's missing, in this ridiculous example, and in the proposed "definition" of fields, is the connection between addition and multiplication, which is, of course, the distributive law. The point is that addition and multiplication do not determine the structure of a field—or do they?

Problem 10 E. *Do there exist two non-isomorphic fields whose additive groups are isomorphic and whose multiplicative groups are isomorphic?*

10 F

10 F. Polynomial values

Can two different polynomials (with, say, rational coefficients) have the same set of zeroes? Sure—that's a trivial question—just multiply a polynomial by 2. All right: can two different monic polynomials have the same set of zeroes? Yes, but that requires a microsecond more of thought. One possible example is

$$x^4 - x^3 \quad (= x^3(x-1))$$

and

$$x^3 - 2x^2 + x \quad (= x(x-1)^2).$$

There is nothing special about the number 0 in questions of this sort: in exactly the same way two different monic polynomials can attain the value -7 at exactly the same set of arguments. Example: add 7 to each polynomial of the example just given.

What happens if two values are considered at the same time?

Problem 10 F. *Can there exist two distinct numbers a and b and two different monic polynomials p and q such that p and q attain the value a at the same places, and, also, p and q attain the value b at the same places?*

10 G. Left inverses

10 G

Can an element of a ring with unity have an inverse that works on one side but not the other? Yes, but we're not likely to run into the phenomenon every day. One example is in the ring of all linear transformations on the vector space of all infinite sequences (of, say, real numbers). If, for instance, U and V are the linear transformations defined by

$$U(x_0, x_1, x_2, \ldots) = (0, x_0, x_1, x_2, \ldots)$$

and

$$V(x_0, x_1, x_2, \ldots) = (x_1, x_2, x_3, \ldots),$$

then $VU = 1$ (= the identity transformation), so that U has the left inverse V, but

$$UV(x_0, x_1, x_2, \ldots) = (0, x_1, x_2, \ldots),$$

so that V is not a right inverse of U. In fact U doesn't have a right inverse at all. Reason: if there were a linear transformation W such that $UW = 1$, then U would have to be surjective, which it is not. (Every vector in the range of U begins with 0.)

Possibilities such as this are the reason why an element x in a ring is called invertible only if there exists an element y in the ring that acts as a two-sided inverse. It is sometimes good to know that a weaker condition implies what is wanted: if x has a left inverse y and a right inverse z, then y must be equal to z and, consequently, their common value is a two-sided inverse. Reason: $yxz = y$, because $xz = 1$, and $yxz = z$, because $yx = 1$.

One consequence of this reasoning is that the number of inverses of an element is either zero or one—if an inverse exists at all, it is unique. For left inverses that consequence is not true. In the U, V example above, for instance, not only is V a left inverse of U, but so also is the peculiar combination $V + U(1 - UV)$. Check:

$$(V + U(1 - UV))U = VU + U^2 - U^2VU.$$

Is that an everyday occurrence or a rare one?

Problem 10 G. *Is there an element in a ring with unity that has exactly two left inverses?*

10 H

10 H. Total product

It is appropriate to open a collection of problems about groups with one that belongs to elementary group theory. The use of that phrase here is almost a pun: some people use the word "elementary" in this context, and in a few other similar ones, not necessarily to refer to the early, easy parts of the theory, but to emphasize that the subject is not groups, and their relations to one another, but the elements of groups.

Problem 10 H. *Can a sequence $\{a_1, a_2, \ldots, a_n\}$ of elements of a group of order n be chosen so that none of the partial products $\prod_{i=p}^{q} a_i$ is equal to the identity?*

Comment. Emphasis: the length of the sequence (which may, of course, contain repetitions) is the same as the order of the group.

10 I

10 I. Cosets

If you are presented with a subset K of an abelian group G, how can you tell whether or not K is a subgroup of G? The question probably sounds a little silly, and the reason is that it is a little silly—surely all that you need to do is to check that the set K is closed under the group operations. (The reason for the plural, "operations", is that inverses have to be checked too.) In other words, K is a subgroup if and only if $x+y \in K$ whenever x and y are elements of K and, at the same time, $-x \in K$ whenever x is an element of K. A quick way of saying all that is just to say that $K - K \subset K$. (The symbol "$K - K$" here is an instance of a notational technique that is frequently useful: whenever A and B are subsets of G, the set of all elements of the form $a - b$, where $a \in A$ and $b \in B$, is denoted by $A - B$. Similarly, of course, $A + B$ denotes the set of all elements of the form $a + b$, where $a \in A$ and $b \in B$.) If that is true, that is, if K is indeed a subgroup of G, then, in fact $K - K = K$; either the inclusion or the equality are characteristic of subgroups.

A subgroup K of the group G determines its cosets, that is, the sets of the form $a + K$. (Note that the last written symbol can be viewed as another instance of the process of forming sums of sets in groups: the coset of G that K associates to an element a of G is the set of all elements of the form $a + k$, where $k \in K$.) If you are presented with a subset S of an abelian group G, how can you tell whether or not S is a coset of some subgroup of G?

Problem 10 I. *Which subsets of an additive abelian group G are cosets in G?*

10 J. Non-trivial automorphisms

The identity map of a group to itself is an automorphism, but surely it deserves to be called the trivial automorphism.

Problem 10 J. *Which groups have a non-trivial automorphism?*

10 K. Automorphisms, $\frac{3}{4}$

The identity automorphism of a cyclic group of order 4 maps exactly half of the elements of the group onto their own inverses. The group of all permutations of the digits 1, 2, and 3 has an automorphism (for example, the inner automorphism induced by the transposition (2 3)) that maps exactly two thirds of the elements of the group onto their own inverses. The numbers $\frac{1}{2}$ and $\frac{2}{3}$ might be the beginning of a simple sequence—what groups is the next term of that sequence associated with?

Problem 10 K. *Is there a finite group with an automorphism that maps exactly three fourths of the elements onto their own inverses?*

10 L

10 L. Automorphisms, $\frac{4}{5}$

Problem 10 L. *Is there a finite group with an automorphism that maps exactly four fifths of the elements onto their own inverses?*

10 M

10 M. Semigroup embedding

A semigroup is a (non-empty) set endowed with an associative multiplication (which, just as in groups, might sometimes be called addition). Every group is a semigroup, but not conversely. The best known semigroup that is not a group is the set of all positive integers (with addition). That example suggests an easy general way to find examples of semigroups: start with a group and throw some of it away. Examples: the set of all odd integers (multiplication); the set of all 2×2 matrices whose determinant is an integer (matrix multiplication).

Not every semigroup comes from a group. An easy example of the other kind is the set of all non-negative integers with multiplication. The trouble is that $0 \times 1 = 0 \times 2$, but $1 \neq 2$; in other words, the cancellation law for multiplication, which is automatically true for any subset of a group, is not true here. Is the failure of the cancellation law the only obstacle?

Problem 10 M. *Can every semigroup with cancellation be embedded in a group?*

Comment. Caution: since semigroups may be non-commutative, there is not one cancellation law, but two.

10 N

10 N. Unions of subgroups

Problem 10 N. *Can a group be the union of two proper subgroups?*

Comment. A proper subgroup is one different from the whole group.

10 O. Maximal subgroups

10 O

A maximal subgroup of a group is by definition a proper subgroup that is not included in any other proper subgroup. Does every finite group have at least one maximal subgroup? The answer is no, because of a degenerate technicality: a group of order 1 has no proper subgroups, and, consequently, can have no maximal subgroups. With that single exception it is true that every finite group has at least one maximal subgroup. Indeed, given a finite group G of order greater than 1, let H_1 be an arbitrary proper subgroup (which always exists, because, for instance, the trivial one-element subgroup is one), and, if H_1 is not maximal, find a strictly larger proper subgroup H_2. (Pertinent comment: it is perfectly possible for H_1 to be maximal; that's what happens if G is a cyclic group of prime order. Note, incidentally, that in this case the group G is abelian.) If H_2 is a maximal subgroup of G, stop; if it is not, find a strictly larger proper subgroup H_3. Keep going this way—since the group is finite, the procedure will lead, sooner or later, to a maximal subgroup. Are the facts for infinite groups the same or different?

Problem 10 O. *Can there be a group that has no maximal subgroups? Can there be an abelian group that has no maximal subgroups?*

10 P. Schröder–Bernstein groups

10 P

The Schröder–Bernstein theorem in set theory says that if there exists a one-to-one mapping from each of two sets into the other, then there exists a one-to-one mapping from either set onto the other. (Actually that theorem says more, but for present purposes this traditional formulation is enough.) Is something like that true in group theory?

Problem 10 P. *If each of two groups is isomorphic to a subgroup of the other, must they be isomorphic?*

CHAPTER 11

SETS

11 A. Lines in the plane

11 A

A line is a small part of the plane, but, obviously, the plane can be covered by lines. How many?

Problem 11 A. *Is the plane the union of countably many lines?*

11 B. Almost disjoint sets

11 B

Finite sets are in a natural sense smaller than infinite sets. Two sets, and, in particular, two infinite sets are called disjoint if their intersection is empty; let us agree to call two sets, and, in particular, two infinite sets almost disjoint if their intersection is finite. A disjoint collection of (non-empty) subsets of a countable set is necessarily countable. (Here "disjoint collection" means "pairwise disjoint", that is a collection of sets such that the intersection of any two of them is empty, and to call the collection countable means just that—that there are only countably many sets in the collection.) Can the statement be strengthened to almost disjoint collections?

Problem 11 B. *Is an almost disjoint collection of (non-empty) subsets of a countable set necessarily countable?*

11 C

11 C. Boundedly almost disjoint sets

What happens to the almost disjoint question if the cardinal numbers of the intersections are required to be bounded? The question has at least two possible interpretations.

One interpretation is this: if a collection of (non-empty) subsets of, say, the positive integers has the property that the intersection of any two of them is a subset of $[1, 1000]$, does it follow that the collection is countable? The answer to that question is yes. For a proof, remove from each set of the given collection those of the numbers in $[1, 1000]$ that it happens to contain; the result is a pairwise disjoint collection of sets. That collection must be countable. Now put back what was removed: each of the (countably many) new sets might have appeared on the scene for many reasons (depending on which elements of $[1, 1000]$ the original set that it came from happened to contain)—but only finitely many reasons. Conclusion: the original collection was countable also.

Another interpretation of the boundedness question suggests a possible strengthening of the Solution 11 B.

Problem 11 C. *If a collection of (non-empty) subsets of a countable set has the property that the intersection of any two of them has not more than* 1000 *elements, does it follow that the collection is countable*?

11 D

11 D. Complex real vector space

An easy question, that is likely to be asked quite early in an introductory course on linear algebra, is whether the set $\mathbb{C}$ of complex numbers is a vector space over the field $\mathbb{R}$ of real numbers. A more pedantically fussy way of asking the question is to replace "is" by "can be made": that is whether it is possible to define a product αz for all real α and all complex z so that the vector space axioms are satisfied. What happens if the roles of $\mathbb{R}$ and $\mathbb{C}$ are interchanged?

Problem 11 D. *Can the additive group* $\mathbb{R}$ *of real numbers be made a vector space over the field* $\mathbb{C}$ *of complex numbers*?

11 E. Rational irrational functions

11 E

Problem 11 E. *Is there a continuous real-valued function on the set of all real numbers whose values are rational on the subset of irrational numbers, and whose values are, at the same time, irrational on the subset of rational numbers?*

11 F. Real semigroups

11 F

The set $\mathbb{R}$ of real numbers can be split into the positive ones and the negative ones; to justify the word "split", the number 0 should be thrown in with one or another set. Each of the two sets so obtained is closed under addition: in technical language, each of them is a semigroup (with respect to addition).

Can $\mathbb{R}$ be split into three additive semigroups? Sure—in a trivial way: use the positive numbers, the negative numbers, and the set consisting of 0 alone. That is not a profound observation; it doesn't yield any new information. What about four?

Problem 11 F. *Can the set of all real numbers be partitioned into four semigroups?*

Comment. The answer offered above for three semigroups is trivial. Does the same question, the one for three semigroups, have a nontrivial answer ? One way to try to find one is to insist that the sets that enter be large—say, for instance, uncountably infinite. That is: can the set of all real numbers be partitioned into three uncountable semigroups?

A semigroup in $\mathbb{R}$ is a set closed under addition; how strongly can that requirement fail? Is there, for instance, an uncountable set of real numbers such that the sum and the product of any two of them is never one of them?

11 G. Closures and complements

11 G

Consider sets in, say, the plane, and look at their complements and their closures. The complement of a set, if you keep forming it, bounces back and forth: starting with A you get A' (a handy and rather common notation for complement), and then A'', which is the same as A, and then

A''', which is A' again, and so on. In technical language (which it is not necessary to use here, but it doesn't hurt), complementation is involutory. On the other hand the closure of a set, if you keep forming it, becomes dull and repetitive immediately: starting with A you get A^- (an unimportant modification of the standard notation that is handy in the present context), and then A^{--}, which is the same as A^-, and then A^{---}, which is A^- again, and so on. In technical language, closure is idempotent.

What happens if we mix closure and complementation? Mixing presumably means that we start with a set and then form closures and complements repeatedly—but, clearly, since a second complement just gets us back where we started from and a second closure doesn't get us any forwarder, it does no good to perform either operation twice in a row. To mix closure and complementation means, therefore, to perform them alternately—but the question about what happens still stands.

Problem 11 G. *What is the largest number of distinct sets that can be obtained from one set by repeated applications of closure and complementation?*

11 H

11 H. Sums of sets

Complex numbers can be added, and so can sets of complex numbers: recall that the sum of the subsets A and B of $\mathbb{C}$ is the set of all numbers of the form $a+b$ with a in A and b in B. This sort of thing can be done in any vector space, or, for that matter in any abelian group. (Non-abelian groups are not illegal, but, as usual, a little more care has to be exercised in dealing with them.)

As preliminary examples, consider the set A of all even integers and the set B of all multiples of 3; what is their sum? (It is, of course, important to specify the sets being added exactly. In the present example *all* integers are under consideration, not only the positive ones. The question makes sense for the positive ones too, but it's a different question with a different sort of answer.) Easy answer: the set of all integers. (Main reason: $(-2)+3=1$.) What if A is the set of all multiples of 7 and B is the set of all multiples of 19? Slightly more difficult number-theoretic answer: the set of all integers again.

It is natural to ask how well behaved set addition is: to what extent it preserves any good properties that the summands may have. The sum of two intervals in the line is an interval; the sum of two closed intervals is a closed interval; is the sum of two closed sets always closed? Answer: no. Example: if A is the set of all integers and B is the set of all integer multiples of $\sqrt{2}$, then $A + B$ is the set of all numbers of the form $p + q\sqrt{2}$, a set that is everywhere dense in the line. (See Problem 2 C.) One conspicuous difference between closed intervals and the sets in this counterexample is that intervals are convex. The sum of two convex sets (in the line, in the plane, or anywhere else where it makes sense) is convex; do the facts about intervals generalize to this situation?

Problem 11 H. *Is the sum of two closed convex sets closed?*

11 I. Cantor plus Cantor

11 I

The sum of two "small" sets is sometimes small (witness the sum of two countable sets, which is always countable), or it may be large. Thus, for instance, the sum of two sets of measure zero might well be very large—witness, the sum of the x-axis and the y-axis in the plane, which is the whole plane. It is pertinent and interesting to ask about one of the most famous sets of measure zero.

Problem 11 I. *If C is the Cantor set, what is the measure of $C + C$?*

11 J. Decreasing balls

11 J

A decreasing sequence of non-empty sets can have an empty intersection. That's obvious; the prototypical example is the sequence $\{I_n\}$ of sets of integers, where $I_n = \{n, n+1, n+2, \ldots\}$.

It is an easy result of general topology that this sort of thing cannot happen with compact sets: the intersection of a decreasing sequence of non-empty compact sets can never be empty. In many friendly metric spaces (in finite-dimensional Euclidean spaces for example) a closed and bounded set must be compact, but the space must be a friendly one; in more general metric spaces it is perfectly possible to have closed and bounded sets that are not compact. Complete metric spaces are not necessarily friendly in this sense, but, nevertheless, they behave in at least a partially friendly manner: in a complete metric space a decreasing sequence of closed balls with diameters tending to zero has a non-empty intersection. What happens if the diameters are not required to tend to zero?

Problem 11 J. *Does a decreasing sequence of closed balls in a complete metric space necessarily have a non-empty intersection?*

11 K

11 K. Decreasing convex

A decreasing sequence of closed and bounded subsets of $\mathbb{R}^2$ has a non-empty intersection. If, however, E_n is the half plane $\{(x, y): x \leqq n\}$ in $\mathbb{R}^2$, then $\{E_n\}$ is a decreasing sequence of closed convex sets with empty intersection: as a substitute for boundedness, intended to guarantee non-empty intersections, convexity is not good enough. If, instead, E_n is the open disk

$$\left\{(x, y): \left(x - \frac{1}{n}\right)^2 + y^2 < \frac{1}{n^2}\right\}$$

in $\mathbb{R}^2$, then $\{E_n\}$ is a decreasing sequence of bounded convex sets with empty intersection: as a substitute for closedness, intended to guarantee non-empty intersections, convexity is not good enough. In general metric spaces, and, in particular, in Banach spaces, even closedness and boundedness together are not enough: a decreasing sequence of closed and bounded sets can have an empty intersection. Can convexity serve as at least a partial substitute for compactness in the infinite-dimensional case?

Problem 11 K. *In a Banach space, does every decreasing sequence of non-empty, closed, bounded, and convex sets have a non-empty intersection?*

CHAPTER 12
SPACES

12 A. Cocountable connected

12 A

Topology has two faces; they used to be called combinatorial and set-theoretic. The terminology has changed; the favored names nowadays are algebraic and general. Algebraic topology is generally regarded as the deeper kind and general topology the broader. What that probably means is that any particular theorem in algebraic topology is harder to come by than the general kind, but the results of general topology are more likely to be usable in many different parts of mathematics. The topological puzzles of this chapter belong to general topology.

The central concept of topology, both algebraic and general, is connectedness. A single point constitutes a connected set, but our intuition is likely to tell us that, with only that exception, connected sets must be quite large, in some sense. Countable sets are small, and, accordingly, our intuition tells us that cocountable sets (the complements of countable sets) should be regarded as large. Are these two uses of the word "large" in harmony?

Problem 12 A. *Must a cocountable set in the plane be connected?*

12 B

12 B. Countable connected

Yes, to be sure, countable sets are small and therefore cocountable sets should be regarded as large; it's no surprise to learn that cocountable sets are usually connected. But what about the "small" sets, the countable ones? Very small sets can indeed be connected—namely the singletons. Are they the only ones?

> **Problem 12 B.** *Can a countable subset of the plane with at least two points be connected?*

12 C

12 C. Decreasing connected

Does a decreasing sequence of connected sets necessarily have a connected intersection? The answer is no; there are easy counterexamples in the plane. Consider, for instance, two parallel rays in the plane, such as the right half of the x-axis and the right half of the horizontal line one unit above it. If C_n $(n = 1, 2, 3, \ldots)$ is the set consisting of those two rays, together with all those points between them whose x-coordinates are greater than or equal to n (the bridge), then $\{C_n\}$ is a decreasing sequence of connected sets whose intersection consists of just the two rays. Could this sort of thing happen with compact sets?

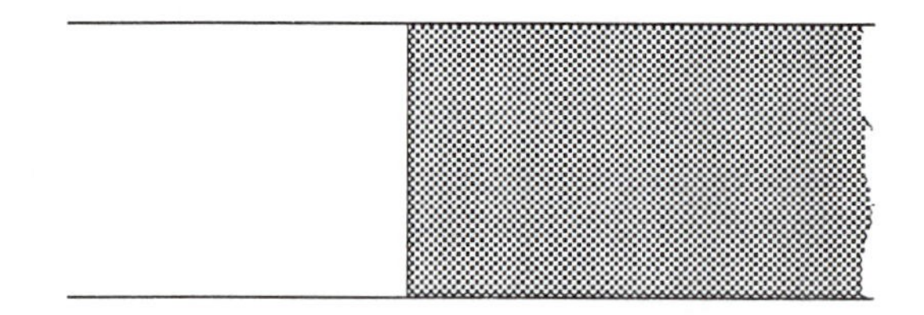

FIGURE 23

> **Problem 12 C.** *Does a decreasing sequence of compact connected non-empty subsets of the plane necessarily have a connected intersection?*

12 D. Plane filling circles, honest

12 D

We all know what the word "circle" means, but there is a temptation sometimes to use the word in a sense different from what we all know. The tempting exceptions are circles of radius 0 (single points) and circles of radius ∞ (straight lines). The plane is the disjoint union of degenerate circles of radius 0 (uncountable union to be sure), and it is also the disjoint union of degenerate circles of radius ∞ (in many ways—one way uses the x-axis and all the lines parallel to it). Let us agree to exclude these degenerate cases and, in this problem anyway, use the word "circle" in its honest meaning only: the circle with center c (a point in the plane) and radius r (a strictly positive number) is the subset of the plane consisting of the points at distance r from the point c. Can honest circles behave as interestingly as degenerate ones?

Problem 12 D. *Is the plane a disjoint union of circles?*

12 E. Plane filling circles, topological

12 E

Is the plane a disjoint union of circles? The question has been asked before and answered (Solution 12 D), but it deserves to be asked again—there is another possible interpretation that demands to be looked at. If "circle" means honest Euclidean circle, the answer is no; but what if "circle" means a topological circle. That's an ad hoc expression; let it be defined, for now, as a subset of the plane that is homeomorphic to an honest circle. So, for example, a topological circle can be (the perimeter of) an ellipse or a rectangle, or, for that matter it can be any simple closed continuous curve, such as the figure below.

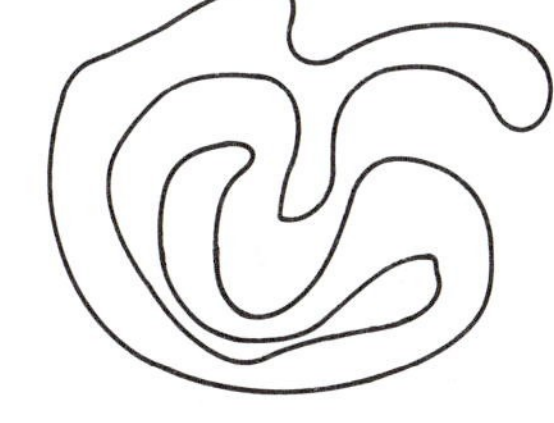

FIGURE 24

Problem 12 E. *Is the plane a disjoint union of topological circles?*

Comment. There is an obvious logical connection between this question and the one in Problem 12 D. If the answer to 12 D had been yes, then the answer to this one would be yes also; and if the answer to this one turns out to be no, then that result would include, supersede, the negative result of 12 D.

12 F

12 F. Space filling circles, topological

There is more room in $\mathbb{R}^3$ than in $\mathbb{R}^2$; even though $\mathbb{R}^2$ is not the union of a disjoint collection of circles, it is perfectly possible that $\mathbb{R}^3$ is. Is it? As before, for the plane, the question has two interpretations: weak or strong, topological circles or honest circles. And, as before, there are implication relations among the possible answers: a no for the topological question implies a no for the honest one, and a yes for the honest question implies a yes for the topological one. What are the facts?

Problem 12 F. *Is $\mathbb{R}^3$ a disjoint union of topological circles?*

12 G

12 G. Space filling circles, honest

The answer to the question of being able to fill $\mathbb{R}^3$ with topological circles turned out to be yes, and that leaves the question for honest circles still open.

Problem 12 G. *Is $\mathbb{R}^3$ a disjoint union of honest circles?*

12 H

12 H. Closed versus open intervals

Trying to express a set as a union of many small and simple sets can sometimes yield some curious geometric information—witness, for example, the representation of $\mathbb{R}^3$ as a disjoint union of circles. A thoroughly non-interesting example is the problem of expressing a closed interval in the line as the union of a collection of open intervals. It cannot be done, and the reason is non-interesting: an end point of a closed interval cannot belong to an open interval that is entirely included in the closed one. (The fact that even disjointness doesn't play a role is another non-interesting feature of the situation.) A small change, however, turns the problem into one to which the answer does not immediately jump to the eye.

Problem 12 H. *Is an open interval a disjoint union of non-degenerate closed intervals?*

Comment. The only thing that the cautionary "non-degenerate" is intended to exclude is singletons.

12 I. Unions of closed intervals

12 I

Is there a higher-dimensional analogue of the result of Problem 12 H? There are several ways of interpreting that question. The following interpretation is among the more interesting ones; the answer to it does not immediately jump to the eye.

Problem 12 I. *Is an open rectangle a disjoint union of non-degenerate closed intervals?*

12 J. Unions of open intervals

12 J

There is at least one more problem about disjoint unions that deserves to be looked at: is a closed rectangle a disjoint union of non-empty open intervals? As it stands the question doesn't seem to have much merit—it is just a 2-dimensional version of the "thoroughly non-interesting" example mentioned in Problem 12 H. The answer is the same here as it was there; the answer is no, because a corner of a closed rectangle cannot belong to an open interval that is entirely included in the rectangle. A change in the question, however, produces a new question that makes a considerable difference between the answers. The change is to replace "honest" open intervals by topological ones. Definition: a topological open interval is a homeomorphic image of an honest (ordinary) open interval.

Problem 12 J. *Is a closed rectangle a disjoint union of non-empty open intervals?*

12 K

12 K. Interval as product

Can you look at a space and tell whether or not it is the Cartesian product of a couple of spaces? Before the question is taken seriously, we should agree that "is" means "is homeomorphic to", and we should exclude from consideration the trivial Cartesian products in which one factor is a singleton.

Problem 12 K. *Is the unit interval a non-trivial Cartesian product?*

12 L

12 L. Convex metrics

A set is convex if along with any two points in it all the points between those two are also in it. The crucial word in this definition is "between". The most natural use of the word is in the theory of ordered sets, but there are others. Other natural interpretations of betweenness, that do not explicitly depend on a concept of order, can be given in vector spaces and in metric spaces. A few aspects of the vector kind were looked at in §5; here is a brief look at the metric kind.

If a, b, and x are points in a metric space X, then the triangle inequality asserts that $d(a,b) \leqq d(a,x) + d(x,b)$. If it happens that the inequality is not strict—if, in other words, equality holds—then there is a geometric sense in which it seems right to say that x is between a and b.

Warning: there might be many points between a and b that have nothing whatever to do with one another. Consider for instance the surface of a sphere in the role of X, with the distance between two points defined as the length of the shortest (great circular) arc that joins them. In that case there are lots of points between the north pole and the south pole—in fact every point other than the two poles is between them.

If the metric on a space X is such that for any two distinct points a and b there exists a point x distinct from both that is between a and b in this sense, then the space is called convex. A sphere is not a convex subset of $\mathbb{R}^3$ in the usual vector space sense, but it is convex in the metric sense just defined. For a different example, let the role of X be played by the union of two distinct great circles through the two poles on the sphere, with the distance that that set inherits from the total sphere. Here too there are many points between the two poles, but the space as

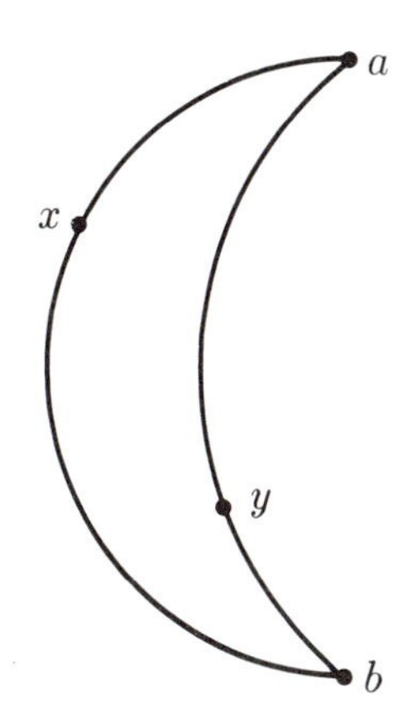

FIGURE 25

a whole is not convex—there are no points between the points x and y indicated in the figure.

The existence of one point between any two distinct points is a pretty weak requirement—it is not at all clear that a metric space satisfying that condition deserves to be called convex. The vector space definition of convexity implies that any two points can be joined by something like an interval; is any shred of that result visible in the metric space theory? If, for instance, the segment between two points a and b in a metric space is defined as the set of all points between a and b, and if a midpoint of a segment is defined as a point x such that $d(a,x) = d(x,b) = \frac{1}{2}d(a,b)$, then it's not at all obvious that midpoints necessarily exist. Do they?

Problem 12 L. *Does every segment in a convex metric space have a midpoint?*

12 M. Circular maps

12 M

Maps, in the sense of geography, are pictures of countries, and the standard rules are that each country must be colored differently from any country with which it has a common border. What's the smallest number of colors needed to be able to achieve that for all maps? It has been known for a long time that three are not enough, and the standard proof of that negative assertion is the map of four countries shown in

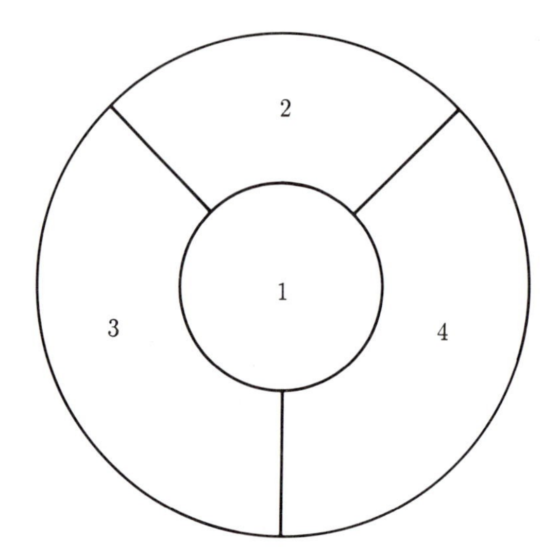

FIGURE 26

Figure 26. Since each of the four countries has a border in common with each of the three others, it is clear that fewer than four colors are not always enough.

The example is simple enough except that the countries it uses do not have simple geometric shapes. Could it be true, for instance, that if

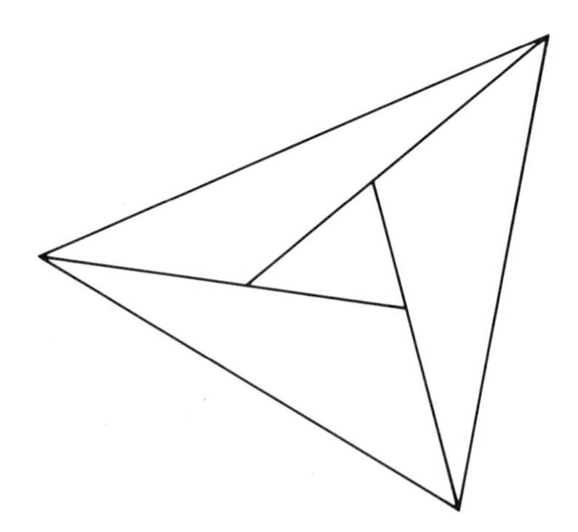

FIGURE 27

all the countries that enter a map are triangles, then fewer than four colors might always suffice? That's a puzzle that requires a little thinking, but not too much. The answer is no: Figure 27 is a map with four countries in which, again, each of the four countries has a border in common with each of the three others, and in which every country is a triangle. What about squares? Could it be that if all the countries that enter a map are squares, then fewer than four colors might always suffice? That might require a little more thought, but the answer remains the same: Figure 28 is a map with six square countries that cannot be colored with fewer than four colors. That needs a tiny bit of argument, as follows. The five outside countries cannot be colored with only two colors, because there are five of them—an odd number—and their colors must alternate. If the outside countries are colored with three colors, then the inside one, which touches each of the outside ones, needs a fourth color.

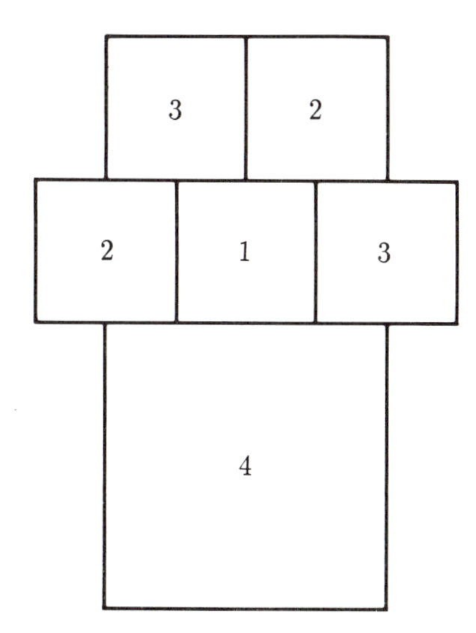

FIGURE 28

It has been known for a short time (since 1976) that if the shapes of the countries satisfy some natural geometric conditions, then four colors always suffice. What if the countries are made up of circles?

Problem 12 M. *A finite number of circles divide the plane into a number of regions and thus define a map in which the border of each country consists of a finite number of arcs of circles. How many colors are needed to color such a map?*

FIGURE 29

CHAPTER 13

MAPPINGS

13 A. Simultaneous fixed points

13 A

The introductory remarks to the chapter concerning spaces pointed out that topology has two faces, the algebraic and the general. There is another classification also, one that pervades every mathematical category, namely the classification into objects and morphisms. In linear algebra the objects are vector spaces and the morphisms are mappings—continuous transformations from one topological space to another. In most categories, and, in particular, in topology, the morphisms are at the center of the stage—their actions are much more important than the static study of the (possibly complicated and possibly beautiful) spaces that they act on—those actions are what topology is really all about. This chapter is devoted to a quick look at a very small number of very special mappings, more to show that there is another world out there than to try to be an exhaustive study.

Problem 13 A. *Must a group of self-homeomorphisms of the closed unit interval have a simultaneous fixed point? Is the answer for abelian groups the same or different?*

13 B

13 B. Maps in maps

Problem 13 B. *Suppose that a rectangle R in the Euclidean plane has on it a map of a country, and suppose that R' is a smaller rectangle that has on it the same map of the same country reduced in the ratio $2 : 1$. Can R' be placed into R in such a way that no point of the country is represented by the same point in R and R'?*

13 C

13 C. Incompressible transformations

A continuous transformation of a topological space into itself can have a very small image—it can, for instance, be a constant, meaning that, for some particular point x_0, it sends every point onto x_0. Such a transformation is, in an obvious sense of the word, a compression—it radically compresses the whole space to a small part of itself. A self-homeomorphism, meaning a homeomorphism of the space with itself, cannot do that, but it can try—it can, for instance, send some sets in the space onto very much smaller proper subsets of themselves. Example: the transformation T defined by $Tx = x^2$ on the set of all non-negative real numbers sends the interval $[0, \frac{1}{2}]$ onto the interval $[0, \frac{1}{4}]$. Since the difference $[0, \frac{1}{2}] - [0, \frac{1}{4}]$ is a substantial set, the transformation T is called compressible. Generally, a self-homeomorphism T is called compressible if there exists a set E such that $TE \subset E$ and $E - TE$ has a non-empty interior; if there is no such set, T is called incompressible. Example: a rotation of a circle (perimeter, that is) is incompressible. How stable (or perhaps better, how algebraic) is the behavior of incompressible transformations?

Problem 13 C. *Is the composition of two incompressible transformations incompressible?*

13 D

13 D. Non-isomorphic topological groups

Problem 10 E pointed out the immorality of considering a set endowed with two different mathematical structures without demanding a structural connection between them. There the focus was on fields, and their additive and multiplicative structures; another, rather different, way that two structures can be anxious to collaborate is visible in the theory of topological groups. A topological group is not just a set that happens to be both a topological space and a group—or is it?

Problem 13 D. *Do there exist two topological groups that are isomorphic as groups and homeomorphic as topological spaces but not isomorphic as topological groups?*

13 E. Connected groups

13 E

The product of a bunch of groups of order 2 is a group in which every element is of order 2. If each factor is topologized with the discrete topology, then the product is naturally topologized with the product topology—which may well not be discrete. Since, however, each factor is compact, the product will be compact for sure (Tychonoff's theorem). But, compact or discrete, it is, for sure, not connected; the sets obtained by fixing any particular coordinate of the product at one of its two possible values are both open and closed. Is the topological phenomenon of disconnectedness a consequence of the algebraic phenomenon of small orders?

Problem 13 E. *Is there a connected topological group in which every element is of order* 2?

Comment. The Cartesian product of two-point spaces is not only disconnected, it is even totally disconnected (in the sense that there is a basis of open-closed sets).

13 F. Inverse mapping theorem for groups

13 F

One of the most celebrated theorems of functional analysis is the inverse mapping theorem. It asserts that a continuous bijective linear transformation between Banach spaces has a continuous inverse. The role of the word "linear" in this context is to emphasize the algebraic structure of the transformation: the theorem says that an algebraically well behaved transformation that is set-theoretically well behaved and is topologically well behaved in one direction must be topologically well behaved in both directions. In this general formulation the theorem makes sense in contexts other than that of Banach spaces—for example, in the context of topological groups. Is it true there?

Problem 13 F. *Is the inverse of a continuous bijective automorphism between topological groups always continuous?*

CHAPTER 14

MEASURES

14 A. Cake problem—wrong

14 A

The traditional way for two children to divide a piece of cake fairly between them is "you cut, I choose". What that process accomplishes is intuitively clear: it cuts the cake C into two pieces C_1 and C_2, with the property that child (1) is happy in the belief that C_1 is worth at least half of C, and, similarly, child (2) is convinced that the value of C_2 is greater than or equal to half of the value of C. Is there a similar procedure that works for three children?

The question is not trivial. Even the two-person question makes non-trivial assumptions that mathematicians should not be willing to take for granted. It tacitly assumes, for instance, that there aren't any tiny, indivisible, parts of the cake (a cherry?, a nut?) that both children place positive value on. The main difficulty of the problem is that one child might like the icing more than other parts (though he likes them too), while, similarly, the other also likes the icing but tends to prefer the filling between two layers of the cake. If they agreed perfectly on how they value different pieces of the cake, and if either one could cut perfectly and perfectly evaluate each part of the cake as worth, in his own estimation, half of the whole, or three quarters, or any other proportion (further assumptions that a mathematically innocent receiver of the problem is probably cheerfully ready to accept), then there would be no problem. In technical language the question is mainly about the existence of certain sets suitably related to two different non-atomic measures defined in the same space, and only partly about an algorithm (what should three children do?).

Suppose that the three children (1), (2), and (3), have associated measures (ways of evaluating pieces of the cake) α_1, α_2, and α_3. Assume, for notational simplicity, that the total value of the cake is equal to 1 in some suitable unit (such as the price that the cake is being sold for). Question: does the cake C have three subsets C_1, C_2, and C_3, that are pairwise disjoint and whose union is C, and that have, moreover, the property that

$$\alpha_1(C_1) \geqq \frac{1}{3}, \quad \alpha_2(C_2) \geqq \frac{1}{3}, \quad \alpha_3(C_3) \geqq \frac{1}{3}?$$

Proposed solution. Let one of the three children, say (1), divide the cake into what he regards as three equal pieces C_1, C_2, and C_3; in other words $\alpha_1(C_1) = \alpha_1(C_2) = \alpha_1(C_3) = \frac{1}{3}$. Let each of (2) and (3) claim dibs on one of those three pieces, the one he prefers the most, the one he would be quite satisfied to have. It is possible that (2) and (3) will claim different pieces, and it is possible that they will claim the same one, but, no matter, at least one of the three pieces C_1, C_2, and C_3 will be left unclaimed. Let (1) remove that piece (one of those pieces) and eat it. So far everybody is satisfied: child (1) got what he believes to be exactly a third, and in the judgment of each of the children (2) and (3) the best piece (the claimed piece) is still available. The problem reduces, therefore, to dividing fairly what's left—and that's the classical problem of dividing a piece of cake (or, who cares, two pieces of cake) between two claimants. Let that problem be solved by "you cut, I choose", and everybody is happy.

Problem 14 A. *The proposed solution is wrong; what's wrong with it?*

14 B

14 B. Cake problem—right

Problem 14 B. *Is there a version of the solution of the cake cutting procedure for two claimants ("you cut, I choose") that works for three claimants?*

14 C. Integrals of variable size

14 C

The value of an integral such as $\int_0^{b(x)} f(t)\,dt$ clearly depends on the upper limit, that is, it is a function of x, and, equally (or even more so!) an integral such as $\int_{a(x)}^{b(x)} f(t)\,dt$ is a function of x. How likely is it that, for non-trivial choices of the parameter functions a and b, the resulting "indefinite" integral is in fact a constant function?

Problem 14 C. *Does there exist a real-valued function f on, say, the ray $(1, \infty)$ such that*

$$\int_x^{x^2} f(t)\,dt = 1$$

for $1 < x < \infty$?

14 D. Measure zero covers

14 D

If E is a Lebesgue measurable set in $\mathbb{R}$, if μ is Lebesgue measure, and if $\varepsilon > 0$, then there exists a sequence $\{I_n\}$ of open intervals such that $E \subset \bigcup_n I_n$ and $\sum_n \mu(I_n) < \mu(E) + \varepsilon$. If $\mu(E) = 0$, can the lengths of the I's be prescribed in advance?

Problem 14 D. *If E is a set of measure zero in $\mathbb{R}$ and $\{\varepsilon_n\}$ is a sequence of positive numbers, does it follow that there exists a sequence $\{I_n\}$ of open intervals such that $E \subset \bigcup_n I_n$ and*

$$\mu(I_n) < \varepsilon_n, \quad n = 1, 2, 3, \ldots ?$$

14 E. Unbounded functions

14 E

A measurable rectangle in the Cartesian product $X \times Y$ of two measure spaces, and, in particular, in $\mathbb{R}^2$, is a set of the form $A \times B$, where A and B are measurable subsets of X and Y respectively. The most natural definition of measurability in product spaces uses the concept of measurable rectangle: according to it a set (in the product space) is measurable if it belongs to the σ-algebra generated by the measurable rectangles. How strongly does the structure of measurable rectangles influence the structure of the generated measurable sets? That's a vague question—here is one way of making it specific: does every measurable set of positive measure (in the product space) include a measurable rectangle of positive measure?

To attack the problem, and to understand the solution, it is advisable to generalize it. The following question (for which it is not obvious that it is indeed a generalization) is pertinent.

Problem 14 E. *Can a measurable function on $\mathbb{R}^2$ be essentially unbounded on every measurable rectangle of positive measure?*

Comment. The word "essentially" describes a property that is unaffected by alterations on a set of measure zero.

Every measurable function, however unbounded it may be, is in a natural way a limit of a sequence of bounded measurable functions. Indeed: given f, consider the functions f_n that are equal to f whenever $|f| < n$ and equal to, say, 0 elsewhere. Consequence: every measurable function is bounded on many large sets of positive measure—an observation that might suggest that the answer to the question in the problem is no.

14 F

14 F. Measures of fixed size

If a measure on the Borel sets of the unit interval frequently agrees with Lebesgue measure, does it follow that it is identical with Lebesgue measure? There are many questions of that sort; here is one of them.

Problem 14 F. *If a measure on the Borel sets of the unit interval takes the value $\frac{1}{2}$ on all sets of Lebesgue measure $\frac{1}{2}$, is it identical with Lebesgue measure?*

14 G

14 G. Intersections of positive measure

Given many large sets, can one always find many among them with a large intersection?

The answer depends, of course, on the meaning one attaches to "many" and "large". The natural meaning of "many" involves cardinal numbers, so that, for instance, a collection could be said to have "many" members if it is uncountable, or if it is just infinite, or even if it is just not empty. If "large" is interpreted to have the same (cardinal number) meaning, then the answer to the question is no. Example: the vertical lines in the plane constitute an uncountable collection of uncountable sets such that the intersection of every subcollection with more than one element is as small as possible, namely empty. Since set-theoretically the plane and the line are the same, it is easy to produce a similar example in the line: there exists an uncountable collection of pairwise disjoint uncountable subsets of, say, the unit interval.

Another possible interpretation of "large" is measure-theoretic. Sample question: does an infinite collection of measurable sets of positive measure in the unit interval always have an infinite subcollection whose intersection has positive measure? This is a trivial question to which the answer is obviously no: just consider an infinite collection of pairwise disjoint intervals (such as $(0, \frac{1}{2})$, $(\frac{1}{2}, \frac{3}{4})$, $(\frac{3}{4}, \frac{7}{8}), \ldots$). A natural way to make the question less trivial is to restrict the values of the measures that are allowed to enter. Example: does an infinite collection of measurable sets of positive measure, with measures bounded away from 0, always have an infinite subcollection whose intersection has positive measure? If the space is an infinite interval (such as $(0, \infty)$ or $(-\infty, +\infty)$), then the answer is no: look at $(0,1)$, $(1,2)$, $(2,3), \ldots$. What makes this example possible? Is it that an infinite interval is an infinite measure space?

Problem 14 G. *Does an infinite collection of measurable sets of positive measure in the unit interval, with measures bounded away from 0, always have an infinite subcollection whose intersection has positive measure?*

14 H. Uncountable intersections

14 H

What made the answer to Problem 14 G negative? The Rademacher sets constitute a countably infinite collection—is that relevant?

Problem 14 H. *Does an uncountable collection of sets of positive measure always have an infinite subcollection whose intersection is a set of positive measure?*

Comment. Note that the question did not assume that the measures of the given sets are bounded away from 0; uncountability makes that assumption unnecessary. The precise statement is that every uncountable collection of sets of positive measure contains an uncountable subcollection with measures bounded away from 0. This is a standard comment. Standard proof: every measurable set of positive measure has measure at least $\frac{1}{n}$ for some positive integer n, and, therefore, if a collection of measurable sets of positive measure is such that, for every n, only countably many of them have measure greater than $\frac{1}{n}$, then the whole collection is countable.

14 I

14 I. Intersections, non-empty

In the present context, contemplation of the curious measure-theoretic behavior of the Rademacher sets leads to asking whether they are at least large enough to have cardinal-theoretically large intersections.

> **Problem 14 I.** *Does an infinite collection of measurable sets in the unit interval, with measures bounded away from 0, always have an infinite subcollection with non-empty intersection?*

HINTS

Chapter 1. Combinatorics

Hint 1 A. The "patient" way to solve the problem is to design an algorithm, based on summing a geometric progression. The "right" way is to note that each match has a loser.

Hint 1 B. If you are one of six people, how many of your friends are friends of one another?

Hint 1 C. If n is even, try to make all the "backward" diagonals constant; if n is odd, look for entries that are missing from the diagonal.

Hint 1 D. What happens if every proper subset of the set of boys knows more girls than is necessary?

Hint 1 E. A first clue to the solution is in the problem obtained by changing 40 to 1. That case, however, is too simple; a good second case is the problem obtained by changing 40 to 2.

Hint 1 F. If the statement is false, then a counterexample must be produced, but if it's true, then infinitely many other questions demand to be asked and answered. Namely: if to every n there corresponds a length k (depending on n, of course, $k = k(n)$), then, for each n, what's the best—meaning, the smallest—value of k? Since the question is about every n, the thinking about it must be inductive.

Hint 1 G. What happens if just two columns are rearranged?

Hint 1 H. Someone shook 8 hands; how many hands did the spouse of that person shake?

Hint 1 I. What does the number 1000 have to do with the question? Does the answer remain the same if 1000 is replaced by any other even number? Any odd number?

Hint 1 J. It's almost impossible to give a hint to the answer that doesn't give the whole show away. It might help, however, to think about some related special questions. Does there exist an infinite sequence of 0's and 1's in which no block of length two occurs three times in succession? What if "length two" is replaced by "length one" or "length three"? What if "three times in succession" is replaced by "twice in succession"?

Hint 1 K. How many sides does a glove have?

Chapter 2. Calculus

Hint 2 A. The very location of the problem (in the section on calculus) is a hint. A slightly more specific hint is that the question is really about limits and that the mean value theorem can be used to good effect.

Hint 2 B. Are there arbitrarily small numbers of the form $\sqrt{n} - \sqrt{m}$? Arbitrarily large ones?

Hint 2 C. How near can a number of the form $p - q\sqrt{2}$ come to an integer?

Hint 2 D. There are computational ways of finding the answer, but thinking is better. Here are some questions that might help the thinking. Does it make any difference whether $s^2 > t$ or $s^2 < t$? In either case, is the sequence $\{x_n\}$ necessarily monotone? What effect does the recursion equation that defines x_{n+1} in terms of x_n have on the possible limit points of the sequence $\{x_n\}$?

Hint 2 E. Are the numbers that enter real or complex? Think about inequalities.

Hint 2 F. Doesn't the geometric series look tempting?

Hint 2 G. Work backward. Start from the equation $(fg)' = f'g'$, and divide by fg. It helps if you recognize a logarithmic derivative when you see one.

Hint 2 H. To work backward is often a good thing to do. Start with the inequality in question and replace x by $\frac{x}{n}$.

Hint 2 I. The question can be regarded as one about trigonometric equations; as such, it is mildly complicated and not particularly elegant, but leads to a precise answer. The question can also be regarded as one about approximation theory; as such, it is simple and elegant, but the answer is not completely trustworthy.

Hint 2 J. Look at the graph not of the function x but of its derivative.

Chapter 3. Puzzles

Hint 3 A. Think not about the part that evaporated, but about the part that cannot evaporate.

Hint 3 B. Is $\sqrt{2}^{\sqrt{2}}$ rational?

Hint 3 C. What's the answer for 30! (instead of 1,000,000!)? If you get the answer six, look again.

Hint 3 D. Obviously the problem can be solved by brute force; is there a pretty solution? Think about eigenvalues, or, better said, about fixed points.

Hint 3 E. Answer: all of them. Be imaginative about permissible operations.

Hint 3 F. Answer: all of them. Observe that two of the three examples that precede the statement of the problem have the same 2×2 top left corner—in other words, the top left corner of a magic square does not uniquely determine it. Question: what additional condition or conditions do determine it?

Hint 3 G. If a number is a sum of two or more consecutive positive integers, ask how many terms the sum has. Start counting from the middle.

Chapter 4. Numbers

Hint 4 A. Is $n^4 + 4^n$ nearly a square?

Hint 4 B. Distinguish between the two cases where the number of 1's is even and odd.

Hint 4 C. Do there exist numbers y and z such that $my \equiv 1 \bmod n$ and $nz \equiv 1 \bmod m$?

Hint 4 D. Think about the Chinese remainder theorem with moduli that are squares of primes.

Hint 4 E. What's the largest possible value that the sum of the digits of some number below a prescribed number can have? For example, if the prescribed number is 434, what's the largest possible value that the sum of the digits of k can be when $k \leqq 434$? A moment's thought should suggest that the answer is equal to the sum of the digits of 399. Incidentally, it helps to remember that every positive integer is congruent modulo 9 to the sum of its digits.

Hint 4 F. How many different odd factors can the numbers between 1 and 100 have?

Hint 4 G. Even without knowing the answer it makes sense to ask how the answer changes as n changes.

Hint 4 H. First guess: all the x's must be about equal. If that's right, so that each x_i is approximately equal to m, say, then there are approximately $\frac{N}{m}$ of them, so that their product is approximately $m^{N/m}$. It's a straightforward calculus problem to determine where the maximum of that occurs. Since it occurs when $m = e$, the answer "must" be approximately $e^{N/e}$. For $N = 10$ that is 39.6—which is not too far from 36.

Hint 4 I. It helps to know that $5^{10} \equiv 1 \bmod 11$.

Hint 4 J. If a power of 3 is less than another power of 3, then the smaller one remains smaller even after it's multiplied by 2.

Hint 4 K. Study what happens with two punches.

Chapter 5. Geometry

Hint 5 A. The pyramid has five faces and the tetrahedron has four, and the gluing removes two faces (one from the pyramid and one from the tetrahedron). Before such reasoning is accepted it is necessary to examine whether the faces form any conspiracies.

Hint 5 B. Given a join closed set, ask how near a point of that set can come to the straight line determined by two points of that set without actually being on the line.

Hint 5 C. If some vertex of the triangle is on the edge of the square but not a vertex of the square, break up the square into two rectangles such that that vertex is a vertex of both of them.

Hint 5 D. This is a minimum problem. What's the best way of attacking it: geometry or calculus?

Hint 5 E. The answer is obviously not four miles (the four sides of the square), nor three miles (omit one of the sides); the two diagonals are better than either of those ($2 \cdot \sqrt{2} = 2.8284\ldots$ miles). Can that be beaten?

Hint 5 F. Can two points an inch and a half apart be joined?

Hint 5 G. What's the maximum number of the prescribed points that can lie on two parallel lines?

Hint 5 H. Is the information about major and minor axes relevant? Does calculus help?

Hint 5 I. Work backward (that is, assume given a triangle that has been bisected), and then repeatedly reflect the picture till you get a hexagon.

Hint 5 J. If there exists a single horizontal line whose intersections with two triangles have equal lengths and cut off equal areas in both, then they are Cavalieri congruent. Try to find such a line by "sweeping" over a triangle with segments from each vertex, one after another, covering the triangle three times (so that the cumulative area goes from 0 to 3), and regard the lengths of those segments as defining a function of the area they cut off.

Hint 5 K. Sometimes the best way to prove that a set is closed is to prove that its complement is open.

Hint 5 L. Consider a natural extreme point (in, say, an interval) and a natural non-extreme point (on, say, a circle), and put them together.

Hint 5 M. Could induction on the dimension possibly work?

Hint 5 N. Don't look for compact counterexamples. Do look for a closed set that is at distance zero from another closed set but, nevertheless, has no point in common with it.

Chapter 6. Tilings

Hint 6 A. The answer is yes—the first player can force a win. A somewhat vague hint that might, nevertheless, be helpful, at least with hindsight, is to think about symmetry.

Hint 6 B. Yes, the first player can always force a win; the forcing move is the simplest one imaginable.

Hint 6 C. Any move that the first player makes enables the second player to reduce the infinite game to a finite one.

Hint 6 D. Try any promising looking forcing move for the first player, and, if it doesn't work, analyze what not working means.

Hint 6 E. If "three" is replaced by "two", then it is easy to see that the answer is yes. Consider, in fact, any two vertices of an equilateral triangle with side length one inch. If they are of the same color, the search is over; if they are of different colors, then the third vertex must have the same color as one of them, and once again the search is over.

Hint 6 F. Tile the plane with squares.

Hint 6 G. Pretend that the originally given large square was a checkerboard. A checkerboard is, to be sure, 8×8, not 10×10, but as far as this problem is concerned that difference is unimportant.

Hint 6 H. Walk through the squares of the checkerboard so that the walk covers every square exactly once and ends at a neighbor of the square where it starts; in other words arrange the squares of the checkerboard in a circular order.

Hint 6 I. Assign coordinates to the squares and examine the sum modulo 3 of all the x-coordinates of the squares of the checkerboard, and also the sum modulo 3 of all the y-coordinates.

Hint 6 J. For checkerboards of size $2^n \times 2^n$, go by induction on the exponent n.

Hint 6 K. Since $\int_a^b e^{2\pi i x}\,dx = 0$ if and only if $b - a$ is an integer, and since

$$\iint_{(a,b)\times(c,d)} e^{2\pi i(x+y)}\,dx\,dy = \int_a^b e^{2\pi i x}\,dx \cdot \int_c^d e^{2\pi i y}\,dy,$$

it follows that the left term vanishes if and only if either $b - a$ or $d - c$ is an integer. No, this is not a misprint, and yes, it has something to do with the question.

Chapter 7. Probability

Hint 7 A. Consider the polynomial of degree 5 whose six coefficients are the p's, and the one whose coefficients are the q's.

Hint 7 B. Try the polynomial factorization technique that worked in Solution 7 A.

Hint 7 C. You *must* put a single dot on some face of both dice, and you *must* put two dots on some face of at least one die. Once that's done, is there any choice left?

Hint 7 D. It's easier to think of renumbering the faces rather than changing their probabilities.

Hint 7 E. Try to construct a set S in the unit square with these properties:

(1) for each t between 0 and 1, the measure of the intersection of S with the vertical strip defined by $0 \leqq x \leqq t$ is less than the measure of the intersection of S with the horizontal strip defined by $0 \leqq y \leqq t$, and

(2) over half the set is below the diagonal $y = x$.

Hint 7 F. The difference between matching and "odd man out" (for three or more people) is gigantic. Matching is decided at the end of the toss, but "odd man out" might not happen on the first toss, or on the second, or, improbable as that seems, on any of the first hundred tosses. Conceivably the game could go on forever without coming to a decision.

Hint 7 G. The probability that I win might be guessed on the basis of a symmetry principle: I am as likely to win as any of my three opponents, no more and no less so, and "therefore" the answer must be $\frac{1}{4}$. Is that reasoning correct? Examine the various different ways that I could win.

Hint 7 H. What about three players?

Hint 7 I. Surely the number $\frac{1}{\pi}$ is irrelevant; what's being asked can and should be asked for any number between 0 and 1.

Hint 7 J. How should the balls be distributed so as to make the probability of drawing a white ball as large as possible?

Hint 7 K. What is the probability that the sum of two random numbers in the unit interval is less than 1? What about three? What about n?

Hint 7 L. It is tempting to say that the repeated drawing of numbers is nonsense—why not just draw one and stop? The chances of winning that way are one in a hundred, and faced with a hundred totally unknown numbers that can't be improved. That's not right. To see that the one chance in a hundred can be improved, think about the following strategy. Draw a number, look at it, discard it, and then keep drawing till you get to a number larger than it was. With that strategy you will surely win if the second number you draw is in fact the largest in the hat (an event of probability $\frac{1}{100}$); but there are many other ways that could lead to a lucky win. Conclusion: the probability of your winning the game with this strategy is greater than $\frac{1}{100}$.

Chapter 8. Analysis

Hint 8 A. The only thing the expression can possibly indicate is a limit—what limit?

Hint 8 B. Express $f(x)$ in two ways as an integral of f', once starting at 0 and once starting at 1.

Hint 8 C. Draw steeper and steeper slopes for less and less time between longer and longer level stretches as you go to the right.

Hint 8 D. Try to compare $\log x$ with powers of $\frac{1}{x}$.

Hint 8 E. Infer integrability from the inequality $\log x \leqq 2\sqrt{x}$, valid for x sufficiently large, and evaluate the integral by using the change of variables that replaces x by $\frac{1}{y}$.

Hint 8 F. The answers are yes, yes, yes, no, and no.

Hint 8 G. Examine sequences such as $\{a, b, c, a, b, c, a, b, \ldots\}$, where $a + b + c = 0$.

Hint 8 H. Construct the horizontal segments that make up the graph of the Cantor function up to the ones of length μ_n, and then join their end points by line segments; the result is a polygonal approximation to the Cantor function whose length is possible to calculate.

Hint 8 I. Assume that f is an entire function with no fixed points and form

$$\frac{f(f(z)) - f(z)}{z - f(z)}.$$

Warning: the proof uses the high powered Picard theorem about the number of values that an entire function can omit.

Hint 8 J. Either integrate $|p|^2$ on the perimeter of the unit circle, or, as an alternative method, examine the behavior of $\frac{p(z)}{z^n}$ at ∞.

Hint 8 K. Given an admissible function f, extend it so as to be defined on and periodic of period 1 on the entire real line.

Hint 8 L. Polynomial in how many variables?

Hint 8 M. If a polynomial (and, in particular, the difference of two polynomials) is uniformly bounded, then it is a constant.

Hint 8 N. A useful lemma is the assertion that the subseries of the harmonic series determined by the square-free numbers diverges. Compare the exponentials (!) of the partial sums of the prime series with the partial sums of the square-free series.

Hint 8 O. What's the answer in case the ε's are all plus 1?

Hint 8 P. Series of the form

$$\sum_{n=1}^{\infty} \frac{\varepsilon_n}{n}$$

can exhibit all possible convergence and divergence behavior. This kind of statement is usually made about series obtained not by changing signs but by rearranging the terms of a conditionally convergent series such as

$$\sum_{n=1}^{\infty} \frac{(-1)^n}{n}.$$

Hint 8 Q. Replace each positive integer in whose decimal representation 0 does not occur by the power of 10 with the same number of digits and estimate the sum of the reciprocals by the sum of the reciprocals of the replacements.

Chapter 9. Matrices

Hint 9 A. If a vector space is the union of a finite number of subspaces, find two vectors: one that belongs to one of the subspaces but not to any of the others, and another that does not belong to the subspace that contained the first.

Hint 9 B. Find as large a subspace as possible that is disjoint from each of a given finite number of subspaces.

Hint 9 C. In this case a small computation is more efficient than thinking.

Hint 9 D. Apply A^{-1} to the Hamilton-Cayley equation that A satisfies.

Hint 9 E. Write any one of the $n+1$ eigenvectors as a linear combination of the others and apply the given transformation to the result.

Hint 9 F. Form A^2 modulo 2.

Hint 9 G. Put A in Jordan canonical form.

Hint 9 H. Let A be the matrix

$$\begin{pmatrix} 0 & 1 & 0 & \cdots & 0 & 0 \\ 0 & 0 & 1 & \cdots & 0 & 0 \\ 0 & 0 & 0 & \cdots & 0 & 0 \\ \vdots & \vdots & \vdots & & \vdots & \vdots \\ 0 & 0 & 0 & \cdots & 0 & 1 \\ \frac{1}{k} & \frac{1}{k} & \frac{1}{k} & \cdots & \frac{1}{k} & \frac{1}{k} \end{pmatrix}$$

and x the vector $(a_0, a_1, \ldots, a_{k-1})$; study the limiting properties of the sequence $A^n x$.

Hint 9 I. Solve the Hamilton-Cayley equation for A in terms of A^2.

Hint 9 J. Infer from the isometry of U that $(x, y) = (Ux, Uy)$ for all x and y, and hence that $U(\alpha x + \beta y) - \alpha Ux - \beta Uy$ is always orthogonal to Uz for all z; use induction on the dimension of the space.

Hint 9 K. Find all the eigenvalues of A.

Hint 9 L. Use Fuglede's theorem.

Hint 9 M. Use the equation $\|T\|^2 = \|T^*T\|$ (true for every linear transformation T).

Hint 9 N. Look for a couple of nilpotent matrices.

Hint 9 O. Let A be a diagonal matrix and B a triangular matrix such that both have distinct diagonal entries.

Hint 9 P. Reduce to the case of positive exponents and use logarithms.

Hint 9 Q. Look for two non-commuting projections.

Chapter 10. Algebra

Hint 10 A. Every real number u, and in particular the putative unity, has a "half"; that is, there exists a number v such that $v + v = u$.

Hint 10 B.
(1) Does every rational number have a half?
(2) How many cube roots of 1 are there?
(3) How many square roots of 1 are there?

Hint 10 C. What role do elementary symmetric functions play as coefficients of polynomials?

Hint 10 D. Consider a polynomial $p(x, y)$ as a polynomial in x with coefficients that are polynomials in y.

Hint 10 E. Consider fields of rational functions in one and two indeterminates.

Hint 10 F. If p and q are polynomials and a and b are distinct numbers, study the relation between the set of zeroes of $(p - a)(q - b)$ and the set of zeroes of $p'(p - q)$ (where p' is the derivative of p).

Hint 10 G. Imitate the behavior of the linear transformation U defined on the vector space of all infinite sequences by $U(x_0, x_1, x_2, \ldots) = (0, x_0, x_1, x_2, \ldots)$.

Hint 10 H. Are all the partial products $\prod_{i=1}^{q} a_i$ different?

Hint 10 I. If a subset S of G is a coset of a subgroup K, then $S-S \subset K$ and $S+K \subset S$.

Hint 10 J. If the group is abelian, consider its elements of order 2.

Hint 10 K. Does the group of all rigid motions of the square work?

Hint 10 L. Possible approach: write $E = \{x\colon Ax = x^{-1}\}$, let x_0 be an element of E, and use a counting argument on $E \cap x_0E$.

Hint 10 M. Form a "free" semigroup and then make it "unfree" by imposing some independent relations that would become dependent if the division were universally possible.

Hint 10 N. If G is the union of H and K, find elements x in H but not in K and y in K but not in H, and ask where xy can be?

Hint 10 O. Consider the additive group G of all dyadic rational numbers modulo 1.

Hint 10 P. Consider an infinite direct product of copies of the additive group of rational numbers.

Chapter 11. Sets

Hint 11 A. How big is a line? This problem is too easy. No reader should be satisfied with only one solution. Three solutions get an A; to get an A+ you need four.

Hint 11 B. For each real number x, consider a sequence of rational numbers that converges to x.

Hint 11 C. Study the mapping that assigns to each set E in the collection the set of 1001 smallest elements in E.

Hint 11 D. The additive group $\mathbb{R}$ is isomorphic to the additive group $\mathbb{C}$.

Hint 11 E. How big can the range of such a function be?

Hint 11 F. Use the isomorphism of $\mathbb{R}^2$ and $\mathbb{R}$.

Hint 11 G. Write $A^+ = A'^-$, and prove that $A^{+++} \subset A^+$. An example that attains the maximum can be found in the line.

Hint 11 H. Look at a hyperbola in the plane.

Hint 11 I. Use the ternary representation of an arbitrary number x in the unit interval to produce two numbers y and z, where y has all the 1's in x, and the 2's in x are split half-and-half between y and z.

Hint 11 J. Try to metrize the set of positive integers so that its tails become closed balls.

Hint 11 K. Look at the space $\mathbb{C}[0, 1]$ of continuous functions on the unit interval.

Chapter 12. Spaces

Hint 12 A. Can two points in a cocountable set be joined by an arc entirely in that cocountable set?

Hint 12 B. Given a countable set C in the plane, find a line that doesn't intersect C but that has points of C on both sides.

Hint 12 C. Assume that the intersection is not connected and use the fact that a compact Hausdorff space is normal.

Hint 12 D. If the interior of a circle of diameter d is a disjoint union of circles, then at least one of them must have diameter less than or equal to $\frac{d}{2}$.

Hint 12 E. If the plane is a disjoint union of topological circles, consider the corresponding disks (where a disk is the union of a circle and its interior), use Zorn's lemma to form a maximal chain of "disks", and then form the intersection of such a chain.

Hint 12 F. Fill the complement of an infinite open cylinder, and then bend that cylinder into a U-shape pushed far away to begin an inductive process.

Hint 12 G. Consider the circles

$$\{(x, y, z) \colon (x - 4k - 1)^2 + y^2 = 1,\ z = 0\}$$

in the (x, y)-plane and their intersections with the spheres

$$\{(x, y, z) \colon x^2 + y^2 + z^2 = r^2\}.$$

Hint 12 H. Assume that the answer is yes and find two sequences of the closed intervals that enter so that the right end points of one converge to the same limit as the left end points of the other.

Hint 12 I. Consider a closed rectangle from the middle of which a closed rectangle was removed and try to express it as a disjoint union of closed intervals.

Hint 12 J. Take care of the corners first.

Hint 12 K. If an interval is a non-trivial Cartesian product, then it is the Cartesian product of two intervals.

Hint 12 L. Given two points, a and b, in a convex metric space X, find a maximal isometry f from $[0, 1]$ to X such that $f(0) = a$ and $f(1) = b$.

Hint 12 M. Ask, for each country, how many circles it is included in?

Chapter 13. Mappings

Hint 13 A. Ask for each transformation in the group whether it is increasing or decreasing.

Hint 13 B. Apply the transformation that "places R' inside R" over and over again.

Hint 13 C. Look at reflections of the line onto itself.

Hint 13 D. Adjoin $\sqrt{2}$ to the rationals to get one group, and adjoin $\sqrt{3}$ to get another.

Hint 13 E. Think about pleasant subsets of the real line being added modulo 2.

Hint 13 F. Form a big direct product, half of which is discrete and half compact.

Chapter 14. Measures

Hint 14 A. Is it possible that after (1) removes a piece, the amount left is less than two thirds in the judgment of both (2) and (3)?

Hint 14 B. Let (1) cut a piece and have (2) and (3) agree that it's fair or else diminish it till it becomes fair.

Hint 14 C. To find out what a solution must look like, work backward: consider a candidate for a solution and differentiate the indefinite integral involved.

Hint 14 D. Look at the Cantor set.

Hint 14 E. The answer is yes, but the construction is not obvious. If χ_n is the characteristic function of the open interval with center r_n and radius $\frac{1}{2n}$, $n = 1, 2, 3, \ldots$, and if

$$k(x, y) = \sum_{n=1}^{\infty} \frac{\chi_n(x-1)}{|x - y - r_n|},$$

then the function k is essentially unbounded on every measurable rectangle of positive measure.

Hint 14 F. Prove that the given measure must agree with Lebesgue measure on all dyadic intervals.

Hint 14 G. Contemplate stochastically independent sets.

Hint 14 H. The separability of the measure algebra implies that every uncountable set in it has a condensation point.

Hint 14 I. Given an infinite sequence of sets with measures bounded away from zero, form their lim sup.

SOLUTIONS

Chapter 1. Combinatorics

Solution 1 A.

1 A

Each match has a loser, and, in fact, there is a one-to-one correspondence between matches and losers. The ultimate champion is the only entrant who doesn't become a loser sooner or later. In other words, the number of losers is 1024, and, therefore, the number of matches is 1024.

Comment. That's pure thought, and we should not have to live without it.

Note that the number 1025, which is $2^{10} + 1$, is a red herring, intended to mislead the innocent—the question can be asked and answered in exactly the same way for any positive integer.

Solution 1 B.

1 B

Yes; $n = 6$ will do. Once that's known, then it follows, of course, that any number greater than 6 will do too. This is an instance of combinatorics in its purest form; it is a non-trivial statement about finite sets that are very finite—six, after all, is pretty small.

For the proof, consider any one of the six points, say A, and the five segments that join it to the others. Three of those segments must have the same color, say red (because $2 + 2 < 5$). Look at the points X, Y, Z that those segments join to A, and consider the triangle with those vertices. If it is a monochromatic blue triangle, we're done. If not, then one of its sides, say $[X, Y]$ is red, and in that case the triangle $[A, X, Y]$ is monochromatic red.

Comment. That's it, and it seems short enough, and it seems elegant enough, but, nevertheless, it is not good enough. Question: now that we know about the existence of at least *one* monochromatic triangle (for six points), what can we say about the possible (or necessary) existence of *two* (or more)? Energetic readers should stop reading right here and try to answer the question; lazy readers can go on to the following paragraph.

How many triangles can six points form? Answer: $\binom{6}{3}$, which is 20 in plain English. If m is the smallest possible number of monochromatic triangles, and b is the largest possible number of bichromatic ones (that means two colors), then clearly $m = 20 - b$; to find m and to find b are the same problem.

A bichromatic triangle has two vertices at which sides of different color meet; call such a pair of sides a *bichrome.* How many bichromes can the segments joining six points form? Five segments issue from each point, and the possibilities for their color distributions are $[5, 0]$ (all five of the same color, no bichromes), or $[4, 1]$ (four of one color and an odd one, four bichromes), or $[3, 2]$ (three of one color and two of the other, six bichromes). The largest possible number of bichromes at any particular point is therefore six. Since the total number of points is six, the number of bichromes altogether cannot be more than 6×6, and since each bichromatic triangle contains exactly two bichromes, it follows that $b \leqq \frac{36}{2} = 18$. Conclusion: $m \geqq 20 - 18 = 2$. In words: for six points there always exist at least two monochromatic triangles.

Is that the best that can be said? Could a sharper proof show that there always exist at least three monochromatic triangles? To answer that question negatively, it is enough to exhibit six points and colored segments joining them so that three monochromatic triangles cannot be found. One way to do that is an easy modification of the five friends-

and-strangers example used earlier. Put six points around a circle (as vertices of a regular hexagon if you like), and then join each neighboring pair with a red segment and all other pairs with a blue one. Draw the figure, look at it, and conclude that there are exactly two monochromatic triangles (blue ones).

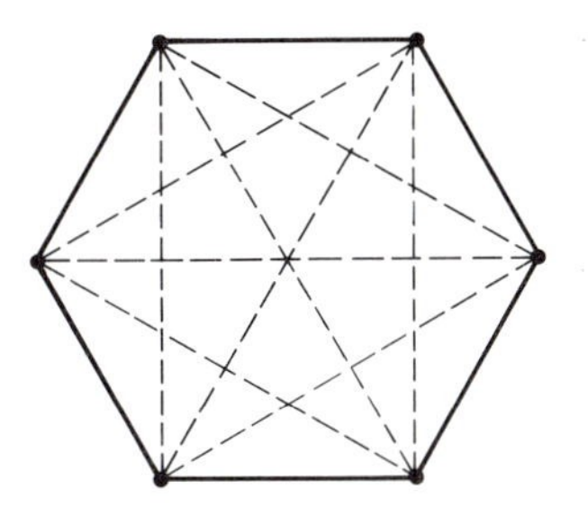

FIGURE 30

The assertion that among any six people there are always three who are all friends or all strangers is known as (the simplest case of) Ramsey's theorem. The general version replaces six by larger numbers, uses more than two colors, and talks about monochromatic subsets with more than three elements; there are also versions that deal with infinite sets. The subject is rich—it seems to be one of the richest parts of combinatorics.

The name Ramsey refers to the most talented member of an unusually talented family. Frank Plumpton Ramsey is internationally recognized and admired as an outstanding logician, economist, philosopher, and mathematician. His father, Arthur Stanley Ramsey, was an applied mathematician who served as president of Magdalen College (pronounced Maudlin) from 1915 to 1937, and his brother Arthur Michael Ramsey, one year younger than Frank, was archbishop of Canterbury from 1961 to 1974. Frank Ramsey made his contributions in a startlingly short time: he died in 1930, not quite 27 years old.

Solution 1 C.

1 C

If n is even, such matrices always exist. For $n = 4$, for instance, here is an example different from the one that appeared before the statement of the problem:

$$\begin{pmatrix} 1 & 2 & 3 & 4 \\ 2 & 3 & 4 & 1 \\ 3 & 4 & 1 & 2 \\ 4 & 1 & 2 & 3 \end{pmatrix},$$

and here is one for $n = 6$:

$$\begin{pmatrix} 1 & 2 & 3 & 4 & 5 & 6 \\ 2 & 3 & 4 & 5 & 6 & 1 \\ 3 & 4 & 5 & 6 & 1 & 2 \\ 4 & 5 & 6 & 1 & 2 & 3 \\ 5 & 6 & 1 & 2 & 3 & 4 \\ 6 & 1 & 2 & 3 & 4 & 5 \end{pmatrix}.$$

These examples suggest the general formula:

$$a_{ij} = i + j - 1,$$

where the addition is to be interpreted modulo n. The pertinent comment is that since $a_{ii} \equiv 2i - 1$ (modulo n), and since n is even, the entries on the principal diagonal are always odd.

If, however, n is odd, then the permutation property of the rows implies that of the diagonal. To establish that statement, it is sufficient to prove that each possible entry $(1, \ldots, n)$ must actually occur in the diagonal. Suppose not—suppose, for instance, that 1 does not. The assumed symmetry of the matrix implies then that 1 occurs an even number of times in the entire matrix—but it occurs exactly once in each of an odd number of rows.

1 D

Solution 1 D.

The conditions listed are sufficient to solve the marriage problem. The proof goes by induction, of course. If the number of boys is just 1, instead of 1000, the result is trivial. The general step, however, from n to $n+1$ contains a slight trap; it would be smart to take a look at the special case of two boys. It looks easy. If the two boys are B_1 and B_2, marry B_1 to a girl he knows, and then ... then there may be a difficulty. Perhaps B_2 had only one girl friend, namely the one that B_1 has already married. In that case the decision to marry B_1 to her was too hasty. Since B_1 and

B_2 between them know at least two girls, B_1 must have an acquaintance different from the one that B_2 knows: marry B_1 to her, and the difficulty has been overcome.

With that preliminary warning, the right way to do the induction step becomes easier to understand. If the number n of boys is greater than 1, and if it happens that every set of k boys, $1 \leqq k < n$, has at least $k+1$ acquaintances, then an arbitrary one of the boys may marry any one of his acquaintances and refer the others to the induction hypothesis. If, on the other hand, some group of k boys, $1 \leqq k < n$, has exactly k acquaintances, then this set of k may be married off by induction. Assertion: the remaining $n - k$ boys satisfy the necessary conditions with respect to the as yet unmarried girls. Indeed, if $1 \leqq h \leqq n - k$, and if some set of h boys were to know fewer than h spinsters, then this set of h bachelors together with the k married men would have known fewer than $k + h$ girls. An application of the induction hypothesis to the $n - k$ bachelors concludes the proof.

Solution 1 E.

1 E

The answer is that the mayor's announcement will be followed by 39 peaceful days and then a massacre of 40 wives before sundown of the fortieth day.

That's not obvious. If, however, the number of unfaithful wives had been just one, Mrs. X, say, then everything would be clear. Indeed, in that case every man except Mr. X would have known about her all along. The mayor's announcement would have given new information to Mr. X only, and he would have shot his wife before sundown that day.

What if there were exactly two unfaithful wives, Mrs. X and Mrs. Y? In that case every man except Mr. X and Mr. Y would have known about two cases of marital infidelity, and each of those two men would have known of just one. As far as Mr. Y is concerned, for instance, the mayor's announcement gives him no new information—but Mr. X's failure to shoot Mrs. X the same day does. The point is that Mr. Y is perfectly able to discover the reasoning of the preceding paragraph, and Mr. X's failure to shoot tells Mr. Y that Mr. X is aware of an unfaithful wife—which could only be Mrs. Y. The same point applies to Mr. X. When the second day after the mayor's announcement dawns, each of Mr. X and Mr. Y comes to be in possession of a proof that his own wife is unfaithful, and two shootings must take place on that day.

An inductive procedure is now running full force: in the case of exactly three unfaithful wives, each of the three men involved would go through the reasoning of the preceding paragraph, and, therefore, remain peaceful for two days and shoot on the third. The inductive passage to 40 is harder to visualize than the passage from 1 to 2, but the logic is inescapable.

Comment. The problem was to prove something about the integer 40, but it's clear—isn't it?—that the spirit of the problem would remain the same if 40 were replaced by any other number. In other words, the problem was really to prove something about every positive integer. It is a basic mathematical principle that the only way to prove something about every positive integer is to use mathematical induction. That has to be true. No matter how the foundations of mathematics are approached, the very meaning of "positive integer" is defined by mathematical induction—that's what the Peano axioms are all about. To apply induction to the problem of marital infidelity is conceptually unusually hard (conceptually, not computationally—what's needed is pure thought, not complicated calculation), and, for that reason, it is one of the best and subtlest examples of how induction can be used.

1 F

Solution 1 F.

The statement is true, and, in fact, it is true that every sequence of n^2+1 (distinct) real numbers has a monotone subsequence of length $n+1$.

For a proof, let the given sequence of numbers be

$$\{a_1, a_2, \ldots, a_{n^2+1}\}.$$

Assign to each integer p (between 1 and n^2+1 inclusive) the length i_p of the longest increasing subsequence that ends with a_p, and, also, assign to each p the length j_p of the longest decreasing subsequence that ends with a_p. Consider any two distinct indices p and q, with $p < q$, say, and ask how it could happen that $i_p = i_q$. Answer: it could happen only if $a_p > a_q$—for otherwise, if a_q were greater than a_p, then a_q could be adjoined to an increasing sequence of length i_p that ends at a_p, and that would yield an increasing sequence of length greater than i_q ($= i_p$) that ends at a_q. Similarly: how could it happen that $j_p = j_q$? Answer (by a completely symmetric reasoning) : only if $a_p < a_q$. Consequence: if p

and q are different, than the pairs (i_p, i_q) and (j_p, j_q) must be different, or, in other words, the correspondence that assigns to each p the ordered pair (i_p, j_p) is one-to-one. Conclusion: the number of ordered pairs such as (i_p, j_p) is n^2+1. If, however, every monotone sequence were of length not more than n, then there would be only n^2 such ordered pairs.

There is a different proof that has some merit. Call a term a_i "major" if it has no successor larger than itself, and call it "minor" if it has no successor smaller than itself. The majors form a decreasing sequence and the minors an increasing one; hence, if the conclusion is false, there can be at most n of each. Remove all majors and minors. The last term a_{n^2+1} is both a major and a minor, so that the number of removed terms is at most $2n-1$, which leaves a sequence of at least $(n-1)^2+1$ terms. By induction, there exists a monotone sequence (say, an increasing one) of length $(n-1)+1$ in it. It follows that the last of the remaining terms must have been a major originally (for otherwise the original sequence would have had a monotone sequence of length $n+1$)—but since all majors had been removed, a contradiction has arrived.

Even when all that is granted, one important question is not answered: is n^2+1 the least length that does the trick? In other words, could it be that every sequence of n^2 (distinct) real numbers has a monotone subsequence of length $n+1$? Or, equivalently but in reverse: could it be that for some values of n there exists a sequence of length n^2 that contains no monotone subsequence of length $n+1$? The answer is yes, and, in fact, it is yes for every value of n. To find such "bad" sequences can be quite a challenge—for small values of n, experimentation will produce them, but even for $n = 3$, the experimentation can be tricky. Here is a general method that works for all n, illustrated for $n = 3$.

Consider the numbers $1, 2, \ldots, n^2$, written as a square array

$$\begin{matrix} 1 & 2 & 3 \\ 4 & 5 & 6. \\ 7 & 8 & 9 \end{matrix}$$

Form the sequence that consists of the first row, backward, followed by the second row, backward, and so on down to the last row, backward; in the present case that becomes

$$3\,2\,1\,6\,5\,4\,9\,8\,7.$$

Any subsequence of length $n+1$ (which is 4 in the present case) must have at least two terms from the same row and must have terms from at least two rows (the reason in both cases is that $n + 1 > n$). Two terms from the same row form a decreasing sequence of length 2 (because they were entered into the sequence backward) and two terms from different rows form an increasing sequence of length 2 (because the order of the rows was not changed). Consequence: any subsequence of length $n + 1$ contains a non-monotone sub-subsequence of length three, and, therefore, no subsequence of length $n + 1$ can be monotone.

1 G

Solution 1 G.

The answer is yes. For the proof, it is sufficient to treat two columns at a time. Given

$$a_1 < b_1$$
$$a_2 < b_2$$
$$a_3 < b_3$$
$$\ldots,$$

rearrange the *rows* so as to make the a's monotone increasing. If the smallest b does not end up in the first row, exchange it with the one in the first row. All pertinent order relations remain true, and now the smallest b *is* in the first row. Induction finishes the argument.

1 H

Solution 1 H.

It is convenient to give each of my fellow guests (my wife included) a name, and a convenient name to give the person who shook k hands is (k), $k = 0, \ldots, 8$. Question: how many hands did the spouse of (8) shake? Answer: none. Reason: (8) and the people whose hands (8) shook exhaust the whole party, except for the spouse of (8), and therefore that spouse is the only one who could have answered "zero". Next: how many hands did the spouse of (7) shake? Answer: exactly one. One way to support that answer is to repeat a modified version of the preceding argument, but there is a better way. Mentally discard the (8)-(0) family from the party (that is: pretend they are not there, and, in particular, do not count any of the hands they shook), and thus reduce the whole problem to a party of 8 instead of a party of 10. In such a party the maximum number of hands any one can shake is 6, and the spouse of the person who shook 6 hands must have shaken none. Consequence:

in the original party, the spouse of (7) must have shaken just one hand (namely that of (8)). A repetition of the argument implies that each couple that I questioned shook exactly 8 hands between them: $8+0$, $7+1$, $6+2$, and $5+3$; that leaves 4 as the only possibility for the number of hands my wife shook.

Comment. A natural first reaction to the question is that it must be a practical joke: the facts given seem to have no relation to any particular person. If, however, you are willing to assume that the problem has a solution, then you can prove, without the analysis given above, that the solution must be 4. The reason is that if exactly the same problem is worded slightly differently, then shake and not-shake are interchangeable. In the different wording it doesn't even matter whether self shakes and spouse shakes are ruled out. The re-worded question that I should ask my fellow guests is "How many hands not in your own family did you shake?" To assume that the problem has a solution means in effect that no matter what symmetric relation (such as hand shaking) I ask about, if the question "How many people not in your own family did you stand in that relation to?" receives nine different answers, then my wife's answer is always uniquely determined and, of course, is always the same, say x. That's true in particular about the relation "shake hands with" and also about the relation "not shake hands with". Since, however, "shake" plus "not-shake" is always equal to 8, so that $x = 8 - x$, it follows that x must be 4.

There is a variation of the handshake problem that might give some people at least a momentary pause, as follows. Suppose that the data you are given is presented not as the result of my questioning, but as a brutal fact: you are told that among the ten people present there are nine who shook different numbers of hands, from 0 to 8 inclusive. Then the tenth person must duplicate one of those numbers—which one?

Solution 1 I.

1 I

The answer is yes: any even number of people can be re-seated so as to preserve their (circular) order and so that no person's number is the same as that of his chair.

Suppose that in the original seating person number $\alpha(1)$ sits in chair number 1, person number $\alpha(2)$ sits in chair number 2, etc.; what is being asserted is the existence of a cyclic permutation γ such that

$\gamma(\alpha(1)) \neq 1$, $\gamma(\alpha(2)) \neq 2$, etc. A cyclic permutation γ of the integers $1, \ldots, n$ is given by

$$\gamma(j) \equiv j + k$$

for some fixed positive integer k, where the addition is interpreted modulo n. The assertion, therefore, is that there exists a value of k such that

$$\alpha(j) + k \not\equiv j \pmod{n}$$

for all j.

How could it happen that there is no such k? That would mean that for each k, there is a value of j (depending on k) such that

$$\alpha(j) + k \equiv j \pmod{n},$$

and, clearly, the value of k could be uniquely recaptured from j. In other words, the transformation that sends each j into its corresponding k would be a permutation, say σ, with the property that

$$(*) \qquad \sigma(j) \equiv j - \alpha(j)$$

for all j. Now sum both sides of the congruence $(*)$ over all values of j. If π is any permutation of the integers modulo n, then

$$\sum_{j=0}^{n-1} \pi(j) \equiv \sum_{j=0}^{n-1} j \equiv \frac{n(n-1)}{2} \not\equiv 0 \pmod{n},$$

the last incongruence being true whenever n is even. (Right? Indeed, if $n = 2m$, then $\frac{n(n-1)}{2} = m(2m-1)$, which cannot be divisible by $2m$.) It follows that the left side of $(*)$ sums to $\frac{n(n-1)}{2}$ and the right side to $0 \pmod{n}$—a contradiction. The non-existence of k is not a tenable assumption, and the proof is complete.

What happens if n is odd? In that case the answer may be no: there exists a "bad" seating, a seating such that in every circular re-ordering of it somebody finds himself in the seat that bears his own number. In other words: there exists a permutation α such that for each k the congruence

$$\alpha(j) + k \equiv j \pmod{n}$$

has a solution for j. The permutation α is easy to describe, but it takes a bit of argument to prove that something with that description indeed exists: $\alpha(j)$ is simply $\frac{j}{2}$ (with the division interpreted modulo n, of course).

What is being asserted is that if n is odd, then every number can be divided by 2 modulo n. Indeed, put $\alpha(1) = \frac{n+1}{2}$, which implies that

$$\alpha(1) + \alpha(1) \equiv 1 \pmod{n},$$

and, for $j \neq 1$, put $\alpha(j) = j\alpha(1) \pmod{n}$, which implies that

$$\alpha(j) + \alpha(j) \equiv j \pmod{n}$$

for every j. Note that the "halving" mapping α is one-to-one (and therefore a permutation) : if $\alpha(i) \equiv \alpha(j)$, then

$$\alpha(i) + \alpha(i) \equiv \alpha(j) + \alpha(j),$$

and therefore $i \equiv j$.

Suppose now that the original seating is the one given by the permutation α. Assertion: in that case there is no k such that

$$\alpha(j) + k \not\equiv j \pmod{n}$$

for all j. Reason: if that non-equation were true, then we could add $\alpha(j)$ to both sides and conclude that

$$k \not\equiv \alpha(j) \pmod{n}$$

for all j. That's a contradiction: it contradicts that α is a permutation. Conclusion: an odd number of people can be seated in circularly arranged chairs numbered from 1 to n in such a way that no matter what cyclic permutation they are subjected to at least one of them will always bear the same number as his chair.

Comment. A purely mathematical formulation of the problem that doesn't involve a "story" is this: for which values of n is it true that corresponding to every permutation α of the integers $\{1, \ldots, n\}$ there exists a cyclic permutation γ such that the composition $\gamma \circ \alpha$ has no fixed points? Answer: exactly the even n's.

1 J

Solution 1 J.

The proposed rule does not work; there does exist an infinite sequence of 0's and 1's in which no finite block occurs three times in succession. It is easy enough to describe such a sequence; the difficulty comes in trying to prove that it works. Write

$$s(n) = 0 \quad \text{whenever } n \equiv 0 \pmod 3,$$

$$s(n) = 1 \quad \text{whenever } n \equiv 2 \pmod 3,$$

and

$$s(3m+1) = s(m) \quad \text{for all } m = 0, 1, 2, \ldots.$$

These conditions unambiguously define an infinite sequence; its first 51 terms (starting at 0 and partitioned into triples for ease of examination) look like this:

001 001 011 001 001 011 001 011 011 001 001 011 001 001 011 001 011.

The first step is easy: no block of length one occurs three times in succession. Reason: each set of three consecutive terms contains both a 0 and a 1, and, accordingly, cannot consist of three 0's or three 1's.

The reason no block of length two occurs three times in succession is that every block of length six must contain a block of the form $0x1$, but neither 010101 nor 101010 does.

Every sequence of length three of non-negative integers contains one that is congruent to 1 mod 3; it follows that every block of length three in the sequence s contains a term of the form $s(3k+1)$. It follows that if a block of length three occurs three times in succession, then, for some k, the terms

$$s(3k+1),\ s(3k+4),\ s(3k+7)$$

all have the same value, and therefore, by the definition of the sequence s, the terms

$$s(k),\ s(k+1),\ s(k+2)$$

all have the same value. In other words, the sequence contains a block of length one three times in succession—and that has already been ruled out.

The next observation is this: if $m > 3$ and m is not a multiple of three, then a block of length m cannot occur three times in succession. Indeed, consider a block of length m, with first term $s(k)$. One of the numbers k, $k+1$, $k+2$ is congruent to 0 modulo 3. If it is $k+j$ (where $j = 0$, or 1, or 2), then $s(k+j) = 0$ (by the definition of s). Since

$$k+m+j \equiv m \quad \text{and} \quad k+2m+j \equiv 2m \pmod 3,$$

it follows that either $k+m+j$ or $k+2m+j$ is congruent to 2 modulo 3. (Reason: m is not a multiple of 3.) Whichever it is, at that position the value of s is 1—which implies that the block

$$s(k),\ s(k+1),\ s(k+2)$$

cannot be the first of three successive occurrences of the same thing.

Suppose, finally, that some block does occur three times in succession and consider one of minimal length. The preceding arguments imply that its length must be $3m$ for some positive integer $m > 1$. The following discussion assumes, for notational convenience, that $m = 3$; the argument, however, will be perfectly general. Every sequence of length 9 ($= 3m$) of non-negative integers contains three terms that are congruent to 1 modulo 3, and it follows, therefore, that the assumed minimal block contains three terms of the form

$$s(3k+1),\ s(3k+4),\ s(3k+7). \tag{1}$$

Since the block is assumed to occur three times in succession, it follows that adding 9 ($= 3m$) to each argument exactly reproduces the same three terms, so that the sequence

$$s(3k+10),\ s(3k+13),\ s(3k+16) \tag{2}$$

is exactly the same as the sequence (1), and the same is true of the sequence

$$s(3k+19),\ s(3k+22),\ s(3k+25). \tag{3}$$

The definition of the sequence s implies that the three short sequences just displayed, all equal to one another, are the same as

$$s(k),\ s(k+1),\ s(k+2),$$

$$s(k+3),\ s(k+4),\ s(k+5),$$
$$s(k+6),\ s(k+7),\ s(k+8)$$

respectively. Consequence: $\{s(k), s(k+1), s(k+2)\}$ is a block of length 3 ($= m$) repeated three times in succession—in contradiction to the assumption that 9 ($= 3m$) is the minimal possible length of such sequences. Conclusion: there is no block that is repeated three times in succession.

Comment. The first solution of the problem of unending chess was offered by Morse and Hedlund. Their sequence is different from the one here presented. It goes like this: start with a 0, and then repeatedly write the complement of everything already written. (The complement of 0 is 1 and the complement of 1 is 0.) The first fifty terms of that sequence, partitioned into quadruples, look like this:

0110 1001 1001 0110 1001 0110 0110 1001 1001 0110 0110 1001 01.

The proof that that sequence works seems to be quite a bit trickier than the proof given above.

It should be pointed out, however, that while both sequences have, at the very least, some mathematical curiosity value, neither of the proffered solutions is really honest. Yes, it is true that infinite sequences of 0's and 1's exist without triply repeated blocks, and the same is true of infinite sequences whose terms are chosen from any desired finite set of symbols. It is true, in particular, that infinite sequences of chess positions exist without triply repeated blocks—but *is there an infinite legal sequence of chess moves without triply repeated blocks*? The point is that not every sequence of chess positions can be achieved by a legal sequence of moves. Given two chess positions, one might not be able to pass from one to the other with a legal move. The existence of an infinite sequence of chess positions without triply repeated blocks does not obviously imply the existence of a sequence with the same property that starts with the standard initial position and that has, moreover, the additional property that *each subsequent term is obtained from its predecessor by a legal move.* Trouble in paradise; I don't know how to make it go away.

Solution 1 K.

1 K

The answer is yes. The point is that four people are to be protected and that two (pairs of) gloves have four available sides. What does the trick is the possibility of turning gloves inside out.

Explicitly: the surgeon puts on both pairs of gloves and performs the first operation. Then he removes, and temporarily puts aside, the outside gloves, and performs the second operation. Finally he turns the first used gloves inside out and puts them on over the ones he just used. This puts into contact two already non-sterile sides—no harm done—keeps his hands in the same sterile situation as they have been all along, and makes available for the third patient the still sterile side (originally the inside) of the first gloves.

Chapter 2. Calculus

Solution 2 A.

2 A

It follows from the mean-value theorem that if u and v are in the interval $[-1, +1]$, then

$$\cos u - \cos v = (-\sin \alpha)(u - v)$$

for some α between u and v, and hence, in particular, for some α in $[-1, +1]$. Since $-\sin 1 \leqq \sin \alpha \leqq \sin 1$, it follows that

$$|\cos u - \cos v| \leqq c|u - v|$$

for all u and v in $[0, 1]$, with $c = \sin 1$. (Note that $|c| < 1$.) Replace u and v by $\cos x$ and $\cos y$ to get

$$|\cos^{(2)} x - \cos^{(2)} y| \leqq c|\cos x - \cos y|$$

for all x and y in $[-\pi, +\pi]$. Do it again: replace x and y by $\cos x$ and $\cos y$ to get

$$|\cos^{(3)} x - \cos^{(3)} y| \leqq c|\cos^{(2)} x - \cos^{(2)} y| \leqq c^2|\cos x - \cos y|$$

for all x and y in $[-\pi, +\pi]$. Apply the same technique inductively to infer that

$$|\cos^{(n+1)} x - \cos^{(n+1)} y| \leqq c^n |\cos x - \cos y|$$

whenever n is a positive integer and x and y are numbers in the domain $[-\pi, +\pi]$.

The problem concerns the sequence $\{\cos^{(n)} x\}$, and in particular its convergence. If it converges, to ξ, say, then $\cos \xi = \xi$. (That's clear, isn't it? If $\xi = \lim_n \cos^{(n)} x$, then $\cos \xi = \lim_n \cos^{(n+1)} x$.) Are there any numbers ξ such that $\cos \xi = \xi$? Sure—just look at the graph of cos, or, for a small bit of highly relevant amusement take a hand-held calculator, set it so that it reads angles in radians, start it with an arbitrary number (1 will do, and so will 0, or $\frac{\pi}{17}$), and then keep pushing the cos button. The result will be $x = .739085\ldots$.

Replace y in the last inequality by ξ to get the result

$$|\cos^{(n+1)} x - \xi| \leqq c^n |\cos x - \xi|$$

for all n and all x. Conclusion: the sequence $\{\cos^{(n+1}(x)\}$ converges to ξ, as n tends to ∞, no matter what x is. Surely that's not obvious when we first look at cosine.

2 B

Solution 2 B.

Since

$$\sqrt{n+1} - \sqrt{n} = \frac{1}{\sqrt{n+1} + \sqrt{n}}$$

(to verify that, cross multiply), it follows that $\sqrt{n+1} - \sqrt{n} \to 0$ as $n \to \infty$, and hence that there are arbitrarily small numbers of the form $\sqrt{n} - \sqrt{m}$. Sample consequence: for some number t of the form $\sqrt{n} - \sqrt{m}$ it is true that $0 < t < .01$.

Since, moreover,

$$k(\sqrt{n} - \sqrt{m}) = \sqrt{k^2 n} - \sqrt{k^2 m},$$

it follows that the set of numbers of that form is closed under multiplication by arbitrary positive integers. Sample consequence: since every positive number is within .01 of one of the numbers t, $2t$, $3t$, $4t, \ldots$, and since all those numbers are of the form $\sqrt{n} - \sqrt{m}$, it follows that every positive number is within .01 of some number of that form.

Replace .01 by any positive number ε (as small as desired), and conclude that every positive number (of whatever size) is within ε of some number of the form $\sqrt{n} - \sqrt{m}$, or, in the appropriate technical phrase, that the set of numbers of that form is everywhere dense in the set of all positive numbers. Consequence: every positive real number is a limit of a sequence of numbers of the form under consideration. Negative numbers cause no difficulty: just interchange n and m. Conclusion: every real number is a limit of the desired kind.

Comment. What happens to the problem, and its solution, if $\sqrt{n}-\sqrt{m}$ is changed to $\sqrt[3]{n} - \sqrt[3]{m}$? Answer: nothing much happens to it. The technique of "rationalizing the denominator" still works, and yields the same conclusion; the only thing that becomes slightly more complicated is the arithmetic involved.

Solution 2 C.

2 C

The answer is that every real number is a limit of numbers of the form $p + q\sqrt{2}$, where p and q are integers, or, in other words, that the set of numbers of that form is everywhere dense.

The method of proof is similar to the one that worked in Solution 2 B. Consider, to begin with, for each integer multiple $q\sqrt{2}$, the nearest integer p just above it, so that $p - q\sqrt{2}$ is between 0 and 1. (The integer p depends, of course, on q.) Given a positive number ε, let n be a positive integer such that $\frac{1}{n} < \varepsilon$, and let q run through the values $1, 2, \ldots, n+1$. The differences $p - q\sqrt{2}$ form a set of $n + 1$ numbers in the interval $(0, 1)$, and, therefore, there must exist two distinct ones among them that are at a distance less than ε from one another. (Look at the intervals $(\frac{k}{n}, \frac{k+1}{n})$.) Subtract them and conclude that there exists a number of the form $p - q\sqrt{2}$ in the interval $(0, \varepsilon)$.

Since the set of numbers of the form under consideration is closed under multiplication by arbitrary integers, it follows that every real number is within ε of one of them; since ε is arbitrary, it follows that the numbers of the form under consideration are indeed everywhere dense.

2 D

Solution 2 D.

Consider a candidate s for a first approximation to $\sqrt{t}$ (where t is a positive number, of course). The case $s = \frac{t}{s}$ is trivial (in that case s is already the desired square root and nothing more has to be done), so that either $s < \frac{t}{s}$ (the guess s is too small), or $\frac{t}{s} < s$ (the guess s is too large). The two possibilities are perfectly symmetric; either can be obtained from the other simply by replacing s by $\frac{t}{s}$. The second possibility is technically a little easier to calculate with; in what follows it will be assumed that $\frac{t}{s} < s$.

If $r = \frac{1}{2}\left(s + \frac{t}{s}\right)$, in other words if r is the average of s and $\frac{t}{s}$, then, of course,

$$\frac{t}{s} < r < s,$$

and therefore $\frac{t}{s} < \frac{t}{r}$.

What is the relation between $\frac{t}{r}$ and r—which is smaller? The answer can be learned from an elementary and not especially messy calculation. Question: is $\frac{t}{r} < r$, or, in other words, is

$$\frac{t}{\frac{1}{2}\left(s + \frac{t}{s}\right)} < \frac{1}{2}\left(s + \frac{t}{s}\right)?$$

Another rephrasing: is

$$4t < s^2 + 2t + \frac{t^2}{s^2}?$$

The last question is equivalent to this: is

$$\left(s - \frac{t}{s}\right)^2 > 0?$$

Since the answer to that is obviously yes, the reasoning can be read from bottom up and proves that the order in which the numbers here studied occur is

$$\frac{t}{s} < \frac{t}{r} < r < s.$$

That's good—that's very good. What it shows is that if for some term x_n of the sequence being examined it is true that

$$\frac{t}{x_n} < x_n,$$

then it follows that

$$\frac{t}{x_n} < \frac{t}{x_{n+1}} < x_{n+1} < x_n.$$

In other words, if it is assumed (without loss of generality) that $\frac{t}{s} < s$, then the sequence $\{\frac{t}{x_n}\}$ is increasing, and bounded above by each term of the sequence $\{x_n\}$, which is decreasing. Consequence: the sequence $\{x_n\}$ converges to a positive limit; call it x.

From the assumed recursion

$$x_{n+1} = \frac{1}{2}\left(x_n + \frac{t}{x_n}\right),$$

it follows that

$$x = \frac{1}{2}\left(x + \frac{t}{x}\right),$$

or, equivalently, that $2x^2 = x^2 + t$. That's it—that's what was wanted: the limit x is the positive square root of t. The answer to the original question is that the sequence $\{x_n\}$ converges for all positive values of s and t and the limit that it converges to is $\sqrt{t}$.

Comment. If you got a B+ or better in calculus and remember Newton's method, the technique just studied is an old friend. What Newton's method says about finding the roots of an equation such as $x^2 - t = 0$ is to guess the answer as, say, x_0, find the tangent line to the curve $y = x^2 - t$ at the point $(x_0, x_0^2 - t)$, and find the intersection of that tangent line with the x-axis. That intersection, call it x_1, is an improved guess. It's an easy exercise in calculus (analytic geometry?) to prove that $x_1 = \dfrac{x_0 + \frac{t}{x_0}}{2}$—Newton's method yields exactly the same approximation method as the one described above.

Solution 2 E. 2 E

There are two different answers, depending on whether the numbers that enter are real or complex. The first two steps in the reasoning are the same in either case:

$$|ax + ay + bx - by| \leqq |a| \cdot |x + y| + |b| \cdot |x - y| \leqq |x + y| + |x - y|.$$

In the real case continue as follows. Since the rightmost (the largest) side of the inequalities just derived is symmetric in x and y, there is no loss of generality in assuming that $|x| \geqq |y|$; since, moreover, that same term is invariant under both the transformations

$$x \to -x \quad \text{and} \quad y \to -y,$$

there is no loss of generality in assuming that $x \geqq y \geqq 0$. After these assumptions are made, the largest side becomes $2x$, which shows that

$$|ax + ay + bx - by| \leqq 2.$$

If

$$a = b = x = y = 1,$$

the left term takes the value 2, which shows that the inequality cannot be improved; the maximum is 2.

In the complex case

$$|x + y| + |x - y| \leqq \sqrt{2} \cdot \sqrt{|x + y|^2 + |x - y|^2}$$

(by the Schwarz inequality, also called the Cauchy–Schwarz inequality, also called the Cauchy–Buniakowsky–Schwarz inequality—the one that says that $|\alpha\xi + \beta\eta|^2 \leqq |\alpha^2 + \beta^2| \cdot |\xi^2 + \eta^2|$—here applied to the case $\alpha = 1, \beta = 1, \xi = |x + y|, \eta = |x - y|$). The last written term is less than or equal to

$$\sqrt{2} \cdot \sqrt{2|x|^2 + 2|y|^2}$$

(by the parallelogram inequality $|\xi + \eta| \leqq |\xi| + |\eta|$), and therefore is less than or equal to

$$\sqrt{2} \cdot \sqrt{4} = 2 \cdot \sqrt{2}.$$

Is that actually equal to the maximum, or could it be improved? To see that it is the maximum, put

$$a = i, \quad x = \frac{1 + i}{\sqrt{2}}, \quad b = 1, \quad y = \frac{1 - i}{\sqrt{2}}.$$

In that case $a \cdot (x+y) = \frac{2i}{\sqrt{2}}$ and $b \cdot (x-y) = \frac{2i}{\sqrt{2}}$, so that

$$|a \cdot (x+y) + b \cdot (x-y)| = \left|\frac{4i}{\sqrt{2}}\right| = \frac{4}{\sqrt{2}} = 2\sqrt{2}.$$

Conclusion: the inequality cannot be improved; the maximum is $2\sqrt{2}$.

Comment. Where does the real argument break down in the complex case?

Solution 2 F.

2 F

Any time something like $\frac{1}{1-r}$ shows up, the temptation is (or should be) strong to set it equal to

$$\sum_{n=0}^{\infty} r^n.$$

Yielding to that temptation, write

$$\frac{1}{1-xy} = 1 + (xy) + (xy)^2 + (xy)^3 + \cdots$$

and then integrate term by term. Since

$$\int_0^1 \int_0^1 (xy)^n \, dx \, dy = \left(\int_0^1 x^n \, dx\right)^2 = \left(\frac{1}{n+1}\right)^2,$$

it follows that

$$\int_0^1 \int_0^1 \frac{1}{1-xy} \, dx \, dy = \sum_{n=0}^{\infty} \left(\frac{1}{n+1}\right)^2 = \sum_{n=1}^{\infty} \frac{1}{n^2}.$$

The last written series is a famous one, and most students sooner or later hear the rumor that it is equal to $\frac{\pi^2}{6} = 1.6449\ldots$.

Solution 2 G.

2 G

To say that $(fg)' = f'g'$ is the same as saying that

$$f'g + fg' = f'g'.$$

Divide that equation through by fg and get

$$\frac{f'}{f} + \frac{g'}{g} = \frac{f'}{f} \cdot \frac{g'}{g}.$$

The worry about the possibility of division by 0 need not be faced at this stage. All that's going on is a formal search for necessary conditions; after some are found they can be examined to see to what extent they are sufficient.

The last equation is the same as

$$\frac{g'}{g} = \frac{-\frac{f'}{f}}{1 - \frac{f'}{f}}.$$

Since fractions such as $\frac{f'}{f}$ are logarithmic derivatives ($\frac{f'}{f} = (\log f)'$), the equation can be rewritten as

$$(\log g)' = \frac{f'}{f' - f}.$$

What this suggests is that f can be prescribed more or less arbitrarily (with a slight courteous bow in the direction of differentiability and the avoidance of division by 0), and, once that's done, g is uniquely determined. Well, not really uniquely: f determines the logarithmic derivative of g, and two more steps are necessary to recapture g itself. First, integrate to get $\log g$; second, exponentiate to get g. The first step introduces an arbitrary additive constant (the one that always enters into indefinite integrals), which the second step converts to a multiplicative constant; in other words, g is uniquely determined by f to within a constant factor.

That much thinking is enough to answer the question: yes, there are many pairs of functions f and g such that $(fg)' = f'g'$.

Comment. Example: set $f(x) = x$, and infer that

$$\log g = \int \frac{dx}{1 - x},$$

so that

$$\log g(x) = -\log(1 - x), \quad \text{or} \quad g(x) = \frac{1}{1 - x}.$$

These calculations may be regarded as mere heuristic hints, but the answer they suggest is checkable. If, indeed, functions f and g are defined on the real line by $f(x) = x$ and $g(x) = \frac{1}{1-x}$ (or, better, on the real line from which the point $x = 1$ has been deleted), then it is a trivial exercise in calculus to check whether $(fg)'$ is or is not equal to $f'g'$, and the answer is that it is.

More generally: set $f(x) = x^a$, where a is an arbitrary real number and compute as above to infer that

$$g(x) = \frac{1}{(a-x)^a}.$$

In case a is negative, the point $x = 0$ must be omitted from the domain of f, and, in any case, the point $x = a$ must be omitted from the domain of g. The verification that $(fg)' = f'g'$ is routine.

A different example: if $f(x) = e^{ax}$, then $\log g = \int \frac{a}{a-1}\, dx$ (and, of course, the value $a = 1$ had better be avoided). Set

$$b = \frac{a}{a-1},$$

infer that $g(x) = e^{bx}$, and verify that once again $(fg)' = f'g'$.

Solution 2 H.

2 H

The answer, which most people find surprising, is that a must be equal to e. For the proof, observe first that values of x less than -1 give no information and cause trouble; ignore them. Given $a^x \geqq 1+x$ whenever $x > -1$, take $x \neq 0$ and raise both sides to the power $\frac{1}{x}$ to get

$$a \geqq (1+x)^{1/x} \text{ if } x > 0 \quad \text{and} \quad a \leqq (1+x)^{1/x} \text{ if } x < 0.$$

Let x tend to 0 and conclude that $a \geqq e$ and $a \leqq e$.

Solution 2 I.

2 I

The natural guess is that the answer is very small, but it is not. An approximation technique, which might be enough to convince most people that the answer is not small, goes as follows. Consider, again, the picture of the situation (Figure 31): where a is a circular arc, whose length is known to be 2640.5 ft, and x is the unknown height, and join the top

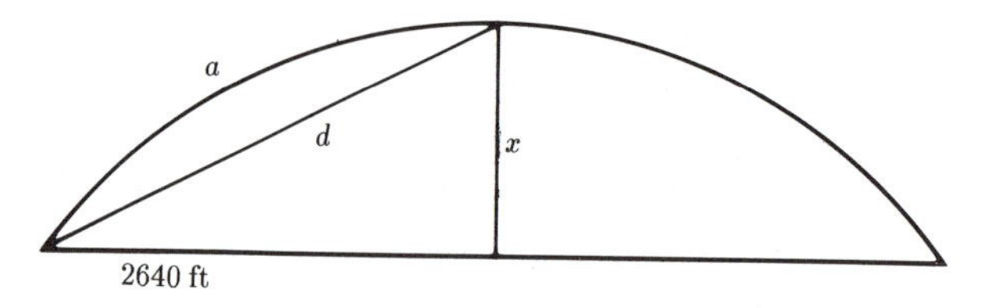

FIGURE 31

point of the interval of unknown length by a line segment to the extreme left point of the railroad track. If d is the length of that segment, then the lengths a and d are not the same, but d looks like a close approximation to a. (Recall that the arc a and the base of the left-hand triangle in the figure differ only by 6 inches.) If a were equal to d, then, by Pythagoras, we would have

$$x = \sqrt{d^2 - (2640)^2} = \sqrt{(2640.5)^2 - (2640)^2} = \sqrt{2640 + .25},$$

so that, approximately, x is (or is more than) the square root of 2500, which is 50. Actually the last indicated square root is equal to 51.38. . . . For this answer properly to be called a part of approximation theory, it would be necessary to study just how far it is from the truth. That sort of thing can frequently be done even when the truth is not known; what is involved is the proof of some pertinent inequalities that show the approximate answer is not farther than something or other from the actual value. Details of that sort would be too far afield here, but even without them the examination has at least some psychological value.

As for the precise truth, consider the detailed picture in Figure 32.

The length A of the arc and the length C of the chord are known; the unknowns are the radius r of the circle of which A is an arc and the angle α subtended by half that arc. The relations among the data and the unknowns imply that

$$r \cdot \sin\alpha = \frac{C}{2} = 2640 \quad \text{and} \quad r \cdot \alpha = \frac{A}{2} = 2640.5.$$

Equivalently:

$$\frac{\sin\alpha}{\alpha} = \frac{C/2}{A/2} = \frac{2640}{2640.5} \quad \text{and} \quad r = \frac{A}{2\alpha} = \frac{2640.5}{\alpha}.$$

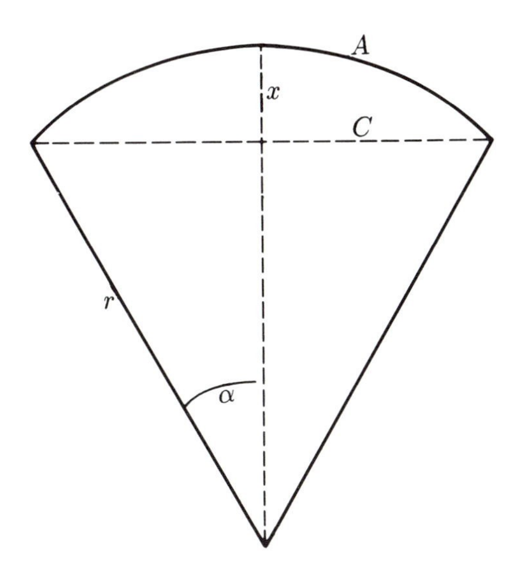

FIGURE 32

To solve these two equations in the two unknowns r and α is not pure joy, but it's not hard. Use $\alpha - \frac{\alpha^3}{6}$ as a pretty good approximation to $\sin \alpha$, and conclude that

$$\alpha = .0337\ldots \text{ (radians)}$$

and

$$r = 78{,}337.31\ldots \text{ (feet)} = 14.84\ldots \text{ (miles)}.$$

Conclusion: $x = r - r \cdot \cos \alpha = 44.497\ldots$ (feet).

Solution 2 J.

2 J

The right way to see the picture is to graph not the function x but its derivative x', and compare it with the triangle whose base is the closed unit interval and whose third vertex is $(\frac{1}{2}, 2)$. The monotoneness of x implies that the graph of x' is always above the t axis; the fact that $x(1) - x(0) = 1$ implies that the area under the graph, like the area of the triangle, is 1. Consequently the graph must cross one of the two equal sides of the triangle at least once; an application of the theorem of the mean shows that the tangent to the graph must be parallel to one of

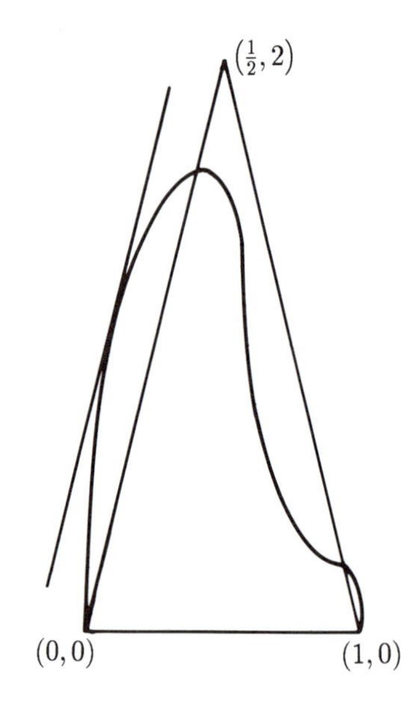

FIGURE 33

those sides at least once. Conclusion: the acceleration must attain the magnitude 4 at least once.

Comment. Can the number 4 in the conclusion be replaced by a larger number? Think about a function x for which the graph of x' is approximately the triangle with a rounded top.

Chapter 3. Puzzles

3 A Solution 3 A.

The answer is 250 pounds. (Were you maybe expecting 490?) Reason: the night before the cucumbers consisted of 5 pounds of solid matter and the rest water. The amount of solid matter stays fixed and, in the morning, it constitutes 2% of the substance that is being asked about. What is the weight that 5 pounds is 2% of (or, in other words, what is 5 pounds multiplied by 50)?

Solution 3 B.

3 B

Is $\sqrt{2}^{\sqrt{2}}$ rational? It seems most unlikely, but no easy argument to prove irrationality is visible. Fortunately for the purpose at hand it doesn't matter whether or not the rationality of $\sqrt{2}^{\sqrt{2}}$ is known; there is a pleasant either–or argument that settles the issue in both cases. All that matters is that $\sqrt{2}^{\sqrt{2}}$ is either rational or irrational. If it is rational, the question is answered (in the affirmative) with $\alpha = \sqrt{2}$ and $\beta = \sqrt{2}$. If it is irrational, then put $\alpha = \sqrt{2}^{\sqrt{2}}$ and $\beta = \sqrt{2}$; the result is that $\alpha^\beta = (\sqrt{2}^{\sqrt{2}})^{\sqrt{2}} = (\sqrt{2})^2 = 2$, and the question is answered (in the affirmative again).

Comment. The mathematics of the argument is trivial, but it belongs to one of the deepest branches of number theory. The difficult problem is to prove that a^b is usually bad when a and b are good. Along those lines the celebrated Gelfond–Schneider theorem (which is a solution of Hilbert's seventh problem) says that if a and b are algebraic numbers with $a \neq 0$, $a \neq 1$, and if b is not a real rational number, then any value of α^β is transcendental (and hence, in particular, irrational). Example: $\sqrt{2}^{\sqrt{2}}$. More complicated and more useful example: since i (the square root of -1) is an algebraic number that is not a real rational number, it follows that if e^π were algebraic, then $e^{\pi i}$ would be transcendental, which it is not—and hence that e^π is transcendental.

Solution 3 C.

3 C

A first idea might be to ask "how many times does 10 go into 1,000,000?", and to use the result, that is 100,000 as the answer. That's obviously wrong. One reason it's wrong is that the product that defines 1,000,000! begins with

$$1 \cdot 2 \cdot 3 \cdot 4 \cdot 5 \cdot 6 \cdots.$$

The point is that the 2 and the 5 form a conspiracy—they contribute a factor 10 that does not occur among the multiples of 10 that the first idea concentrated on.

All right: a corrected form of the question is "how many times does 5 go into 1,000,000, and how many times does 2?" On second thought, surely it is not necessary to ask about 2. There are lots and lots of 2's available—more than enough to match however many 5's we end up with. So: "how many times does 5 go into 1,000,000?". Answer: 200,000. Is that the right question to ask, and is the answer to that question the answer to the main question?

No, but this time the reason why it's no is a little subtler. The reason is that "go into" is the wrong question to ask. Think, for instance, about the question of how many zeroes 10! ends in. Asking how many times 10 goes into 10 gives the wrong answer (one), and asking how many times 5 goes into 10 gives the right answer (two); but even the second question is the wrong one to ask. Indeed: what about 11!? Since neither 10 nor 5 "goes into" 11, the method yields the answer that 11! ends in no zeroes at all—which is silly. Surely in passing from 10! to 11! no zeroes are lost.

A modified question, that avoids the error of paying too much attention to the zeroes that are already present in the given number (as in 10 and as in 1,000,000) is this: "how many multiples of 5 are there below 1,000,000?" (It's too much trouble to say "below or equal to"—let "below" be understood to include the case of equality.) The answer is 200,000, and that's the same as the answer for 1,000,001—a good sign. But the question is still the wrong auxiliary question to ask, and the reason it's wrong can be seen by examining a number such as 30 (or, for that matter, 31) in place of 1,000,000. The answer to "how many multiples of 5 are there below 31?" is six, but 31! ends in seven zeroes, not six. The point is that the product that defines 31! contains the six factors

$$5,\ 10,\ 15,\ 20,\ 25,\ 30,$$

one of which, namely 25, contributes two 5's (and therefore two zeroes at the end).

The truly right question to ask is this: "how many 5's occur in the prime factorization of all the positive integers below 1,000,000?" That question is a compressed way of asking several questions. There are

$$\begin{aligned}
200{,}000 &\text{ multiples of 5 below } 1{,}000{,}000,\\
40{,}000 &\text{ multiples of } 5^2,\\
8{,}000 &\text{ multiples of } 5^3,\\
1{,}600 &\text{ multiples of } 5^4,\\
320 &\text{ multiples of } 5^5, \text{ and}
\end{aligned}$$

$$64 \text{ multiples of } 5^6.$$

Tempting conclusion: since 5^7 does not divide 1,000,000, the time has come to stop and add up the results obtained. The sum is 249,984—isn't that the number of zeroes that 1,000,000! ends in?

No, it is not; one more, final, correction is needed. The fact that 5^7 does not divide 1,000,000 is true but irrelevant. The search for the zeroes that 1,000,001! ends in would have proceeded exactly the same way up to this point, even though 1,000,001 is not divisible by any multiple of 5; divisibility is still the wrong feature to emphasize. The displayed list should go on with powers of 5 as long as multiples of them below 1,000,000 can be found, thus:

$$12 \text{ multiples of } 5^8,$$

$$2 \text{ multiples of } 5^9.$$

Now the counting can stop: 5^{10} is greater than 1,000,000. To learn that there are 12 multiples of 5^8 below 1,000,000 just divide:

$$\frac{1{,}000{,}000}{78{,}125} = 12.8,$$

and note that the fractional part contributes nothing to the count.

Final answer: 249,998 $(= 249{,}984 + 14)$.

Comment. The same question can be asked even if the notation for integers uses some base other than 10. Whatever base b is used, the symbol 1,000,000 indicates b^6. So, for instance, if the base is 2, then one million is 2^6, which is what we usually call 64; the method used in the solution above yields the result that 1,000,000! ends in 63 zeroes. What if the base is 7? or 6? or 12?

Solution 3 D.

3 D

Write $T(x)$ for the number of coconuts that remain from a pile of x coconuts after a typical sailor's operation; clearly $T(x) = \frac{4}{5}(x-1)$. Call an integer x a "solution" if the result of starting with x and iterating T six times is an integer; the problem is to find the least positive solution.

Since $T(x) = \frac{4}{5}x - \frac{4}{5}$, so that

$$T^6(x) = \left(\frac{4}{5}\right)^6 x - \left(\left(\frac{4}{5}\right)^6 + \left(\frac{4}{5}\right)^5 + \cdots + \frac{4}{5}\right),$$

it follows that

$$T^6(x) - T^6(y) = \left(\frac{4}{5}\right)^6 (x - y)$$

for all x and y. This implies that if x and y are solutions, then x and y are congruent modulo 5^6, and, conversely, if x is a solution and x and y are congruent modulo 5^6 then y is a solution.

Up to here the reasoning looks as if it were leading up to a brute force calculation, but at this point an inspiration becomes possible. The inspiration is to ask whether the operation T has any "eigenvectors", or, more properly speaking, any fixed points. Never mind why anyone should ask that; let's see what the answer is. Could it happen that $T(x) = x$? The question is a trivial linear equation, one equation in one unknown, and the unique answer is obviously $x = -4$, which seems to make no sense in this context. Formally, however, -4 is a solution (note that the definition was phrased so as not to exclude negative integers), and, therefore, so is every other number that is congruent to -4 modulo 5^6. The smallest positive solution now becomes obvious: it must be $-4 + 5^6$, or, in other words, 15,621.

Comment. Why would anyone think to ask about "eigenvectors"? Folklore attributes this solution to the great mathematical physicist P. A. M. Dirac—who was used to solving problems by using eigenvectors. It is possible that his active vocabulary did not contain the expression "fixed point", and that, therefore, while he was thinking of the right thing, he used a familiar but slightly off-center phrase to refer to his thoughts. Except for the curiously ingenious solution here described, the problem has no mathematical merit.

3 E

Solution 3 E.

If the function $\sqrt{\ }$ (square root) of one variable and the function log of two variables ($\log_a b$) are allowed, then every positive integer can be expressed in terms of three 2's. Consider, as a typical example, the expression

$$\log_2\left(\log_2\sqrt{\sqrt{\sqrt{\sqrt{\sqrt{2}}}}}\right).$$

Clearly it contains exactly three 2's; what is its value? If the tower of square roots is expressed in terms of fractional exponents, the whole becomes

$$\log_2\log_2(2)^{1/2^5} = \log_2\left(\frac{1}{2^5}\right) = \log_2 2^{-5} = -5,$$

so that one more step (subtract from 0, or, simply said, change the sign) yields 5. If the number of times that the square root is used is not five, but 99, the result becomes 99; the idea works in complete generality.

Comment. So far as I know this mathematical bauble, something to do while you are listening to a lecture you don't want to listen to, was discovered by John von Neumann.

Solution 3 F.

3 F

Every $(n-1)\times(n-1)$ matrix can be the top left corner of an $n\times n$ magic square. If, in fact, an arbitrary $(n-1)\times(n-1)$ matrix is prescribed, together with an arbitrary number that is to be the common value of the row sums and the column sums, then there is a unique magic square with the prescribed top left corner and the prescribed sums. That may be an unexpected result, but, once it is faced, it causes no difficulties—its proof is easy, and is uniquely forced by its statement. The case $n=3$ is completely typical; nothing is lost by restricting the proof to that case.

Suppose then that it is desired to construct a magic square whose top left corner is

$$\begin{pmatrix} a & b \\ c & d \end{pmatrix}$$

and whose row and column sums are equal to s. In other words, it is desired to determine x, y, u, v, and Q so that

$$\begin{pmatrix} a & b & x \\ c & d & y \\ u & v & Q \end{pmatrix}$$

is a magic square with row and column sums equal to s. There is obviously no choice—x, y, u, and v must be determined so that

$$a + b + x = s \qquad c + d + y = s$$

$$a + c + u = s \qquad b + d + v = s,$$

and those simple equations can be uniquely solved for x, y, u, v. The only possible worry is whether Q can be found—the difficulty is that Q must satisfy two conditions, not one, and, for all we know, they are not compatible. The question is whether $x + y$ and $u + v$ are equal or not. The answer can be found by looking: is

$$(s - (a + b)) + (s - (c + d))$$

equal to

$$(s - (a + c)) + (s - (b + d))?$$

Since the answer is yes, there is no problem, and the proof is complete.

Comment. How large a subset of the set of all $n \times n$ matrices is the set of all $n \times n$ magic squares? One way to understand the question is to note that the set of all $n \times n$ matrices with entries that are real numbers (or, for that matter, elements of any field) is a vector space of dimension n^2. The set of all $n \times n$ magic squares is a subspace of that vector space; one natural way to interpret the question about size is as a question about dimension. The preceding solution answers that question. Indeed, an $n \times n$ magic square is uniquely specified by prescribing its $(n-1) \times (n-1)$ top left corner and prescribing the sum of its first row, and they can be prescribed arbitrarily. Conclusion: the set of all $n \times n$ magic squares is a vector space of dimension $(n - 1)^2 + 1$. So, for instance, the set of all 3×3 magic squares is a subspace of dimension 5 of the vector space of all 3×3 matrices, a vector space of dimension 9.

A similar question can be asked and answered about magic squares with integer entries in exactly the same way. The pertinent concept is that of the rank of an abelian group. The basic properties of that concept are the same as the properties of dimension. The comment above was made about dimension and fields, rather than about rank and groups, because dimension is usually regarded as more elementary and easier than rank.

Solution 3 G.

3 G

If n is the sum of two or more consecutive positive integers, ask how many terms the sum has. If it has an odd number of terms, say $2k + 1$, then consider the middle term, say m, and look to both sides from the middle:

$$(m - k) + \cdots + (m - 1) + m + (m + 1) + \cdots + (m + k).$$

When these terms are added, a lot of cancellation takes place, and the sum, known to be n, is seen to be equal to $(2k + 1)m$.

What happens if n is the sum of an even number, say $2k$, of consecutive positive integers? If the sum is written in the increasing order of its terms, then there is no middle term, of course—but there are an odd number of plus signs between the terms, and there is a middle one among them. If, for instance, $n = 44$, and the sum is

$$2 + 3 + 4 + 5 \,[+]\, 6 + 7 + 8 + 9,$$

with the middle plus sign indicated by the brackets, then the pairs of numbers equally distant from it always have the sum 11 (which makes it obvious that the entire sum is indeed 44). The general even case looks just like the example:

$$(m - k + 1) + \cdots + (m - 1) + m \,[+]\, (m + 1) + (m + 2) + \cdots + (m + k);$$

pairs of terms equally distant from the center always have the same sum, namely $2m+1$, and therefore the sum, known to be n, is seen to be equal to $k(2m + 1)$.

In either case, odd or even, n always has an odd factor greater than 1: the factor $2k + 1$ in the one case and the factor $2m + 1$ in the other. Which numbers do *not* have an odd factor greater than 1? Answer: the powers of 2. Conclusion: a necessary condition that n be the sum of two or more consecutive positive integers is that it is not a power of 2.

The argument looks completely reversible. If $n = (2p+1)q$ (it's a good idea to use letters different from the ones above, so as not to be prejudiced whether we are in the odd case or the even), then

$$n = (q-p) + \cdots + (q-1) + q + (q+1) + \cdots + (q+p).$$

Alternatively, thinking of n as $q(2p+1)$, we could write

$$n = (p-q+1) + \cdots + (p-1) + p\,[+]\,(p+1) + (p+2) + \cdots + (p+q),$$

and thus conclude, again, that n is the sum of two or more consecutive positive integers.

Does that look right? The answer is that it is nearly right, but there is something fishy about it. The reversed argument seems to indicate that no matter what n is, so long as it has an odd factor greater than 1, it can always be written as the kind of sum wanted, with the number of terms being either odd or even, whichever is wanted. Simple examples show that that is false: $5 = 2+3$, but no odd number of terms will work, and $6 = 1+2+3$, but no even number of terms will work. What's wrong?

What may be wrong is that the representations of n may contain terms that are not positive. If it happens that $q < p$, then $q-p$ is negative, and that rules out the first representation; if it happens that $p < q$, then $p-q+1$ is zero or negative, and that makes the second representation either suspicious or illegal. Nevertheless, with just a little extra care, the technique works. The simplest way to make it work is to look at $n = (2p+1)q$, and see which is greater, p or q; use the first representation if $q > p$ and the second if $p \geqq q$.

Comment. Actually in most cases both representations work; the difficulty is more illusory than real. If, for instance, $p = 3$ and $q = 2$, so that $n = 14$, then the first representation is

$$14 = (-1) + 0 + 1 + 2 + 3 + 4 + 5,$$

which is illegal, to be sure—but if the -1 is allowed to cancel the $+1$ and the 0 is just peacefully discarded, the result is the valid representation

$$14 = 2 + 3 + 4 + 5.$$

Similarly, if $p = 2$ and $q = 5$, so that $n = 25$, then the second representation is

$$25 = (-2) + (-1) + 0 + 1 + 2 + 3 + 4 + 5 + 6 + 7,$$

which is illegal. Let the negative terms cancel what they can, and

$$25 = 3 + 4 + 5 + 6 + 7$$

is left, which is just fine.

One of the sources of beauty in mathematics is the number of ways changes can be rung on old questions to yield new ones. Here is a sample: which positive integers are sums of two or more consecutive *odd* positive integers? (A necessary and sufficient condition that a number be so representable is that it be either a multiple of 4 or else a composite odd number.) Second sample: Which positive integers are sums of *three* or more consecutive positive integers? (A necessary and sufficient condition that a number be so representable is that it not be either a power of 2 or a prime.)

Chapter 4. Numbers

Solution 4 A.

4 A

If n is even, then $n^4 + 4^n$ is even and greater than 2, and, therefore, it is not a prime. If $n^4 + 4^n$ could be factored when n is odd, then, once again, the problem would be solved, but no factoring seems to lie near the surface. There is one, however, an inch below the surface. The trick is to write

$$n^4 + 4^n = (n^2 + 2^n)^2 - 2^{n+1}n^2.$$

If n is odd, so that $\frac{n+1}{2}$ is an integer, then the right side above factors as

$$(n^2 + 2^n - 2^{(n+1)/2}n) \cdot (n^2 + 2^n + 2^{(n+1)/2}n),$$

and that looks as if it settled the matter: it seems to say that the expression in question can be a prime only when $n = 1$. But there is a catch.

It is a small catch and it's easy to fix, but it shouldn't be forgotten:

it must be proved that for $n = 3, 5, 7, \ldots$ neither of the factors of n^4+4^n so exhibited is equal to 1. A possible proof is a small calculation: notice that

$$n^2 + 2^n - 2^{(n+1)/2}n = (n^2 + 2^n - 2 \cdot 2^{n/2}n) + (2 \cdot 2^{n/2}n - 2^{(n+1)/2}n).$$

The first summand on the right is a square, and the second summand is a monotone increasing function of n. That second summand is less than 1 when $n = 1$, but it becomes greater than 1 as soon as n reaches 2, and, therefore, it stays greater than 1 forever after.

4 B

Solution 4 B.

If N is one of the numbers in question in which the number of 1's is k—if, for instance, $k = 4$ and

$$N = 1010101,$$

then the product $11 \cdot N$ has $2k$ digits all equal to 1, as in

$$11 \cdot 1010101 = 11111111 = 1111 \cdot 10^4 + 1111 = 1111 \cdot (10^4 + 1).$$

This kind of factoring always works; it is always true that

$$11 \cdot N = M \cdot 10^k + M = M \cdot (10^k + 1),$$

where M is the number that has k digits all of which are equal to 1.

To finish the solution, it is useful to know the relation between M and 11 as well as the relation between $10^k + 1$ and 11. It is pretty clear that if k is even, then M is divisible by 11, and if k is odd then M is not divisible by 11. For $10^k + 1$ it's the other way around: if k is odd then it is divisible by 11, and if k is even, then it is not, but the reasoning is slightly subtler. For the odd case the value $k = 5$ is typical:

$$10^5 + 1 = 100001 = 111111 - 10 \cdot 1111,$$

which implies divisibility by 11. For the even case, such as

$$10001 \; (= 10^4 + 1),$$

observe that the number differs from 9999 by 2—which implies non-divisibility by 11.

The conclusion is now accessible. Whether k is even or odd, exactly one of M and 10^k+1 is divisible by 11, and, since the product of M and $10^k + 1$ is equal to $11 \cdot N$, it follows that N is divisible by the other one. If $k = 2$, then $M = 11$ and $10^k + 1 = 101$; the result is that $11 \cdot 101 = 11 \cdot 101$, which is neither inspiring nor informative. In every other case, the reasoning just used implies that N is divisible either by 11 or by $10^k + 1$, neither of which is equal to 1 or to N.

Final conclusion: the only N that can be prime is 101.

Solution 4 C.

4 C

The general solution is a faithful copy of the one for the special case

$$x \equiv 5 \bmod 11,$$

$$x \equiv 7 \bmod 17,$$

and numbers are sometimes easier to keep straight than the alphabet.

A simpler, and helpful, question is this: do the simultaneous congruences

$$y \equiv 0 \bmod 11,$$

$$y \equiv 1 \bmod 17,$$

have a solution? That is, can a multiple of 11 be congruent to 1 modulo 17? The answer is not obvious from scratch, but it is obvious to anybody who has been exposed to the elementary facts about greatest common divisors. Indeed, since the greatest common divisor of 11 and 17 is 1, it follows that there exist integers y and z such that

$$17y + 11z = 1.$$

From there the answer springs to the eye: since $11z = 1 - 17y$, it is indeed the case that a multiple of 11 can be congruent to 1 modulo 17.

A similar simpler question could have been asked with the roles of 11 and 17 interchanged: do the simultaneous congruences

$$z \equiv 1 \bmod 11,$$

$$z \equiv 0 \bmod 17,$$

have a solution? The same fact about greatest common divisors answers that one too: since $17y = 1 - 11z$, it is indeed the case that a multiple of 17 can be congruent to 1 modulo 11.

The thinking is now finished; all that follows is arithmetic. Multiply the congruences involving y by 7 to get

$$7y \equiv 0 \bmod 11,$$

$$7y \equiv 7 \bmod 17,$$

and, similarly, multiply the congruences involving z by 5 to get

$$5z \equiv 5 \bmod 11,$$

$$5z \equiv 0 \bmod 17.$$

What about the sum of these results: $7y + 5z$? Miracle: modulo 11 it is $0 + 5$, and modulo 17 it is $7 + 0$—victory!

Can the actual numbers be obtained by such techniques? Sure: all that needs work is the greatest common divisor equation, $17y+11z = 1$, and that can be found by inspection: $y = 2$ and $z = -3$. In other words:

$$17 \cdot 2 + 11 \cdot (-3) = 1,$$

so that

$$17 \cdot 2 \equiv 1 \bmod 11 \quad \text{and}$$

$$11 \cdot (-3) \equiv 1 \bmod 17.$$

Consequence: the number

$$5 \cdot 17 \cdot 2 + 7 \cdot 11 \cdot (-3)$$

does everything that is wanted of it: modulo 11 it is 5 and modulo 17 it is 7. The number so obtained is in fact $170 - 231 = -61$. Modulo 11 that is -6 (which is 5) and modulo 17 that is -10 (which is 7). If positive solutions are preferred, just add something that is congruent to 0 modulo both 11 and 17—that is add $11 \cdot 17 = 187$, and thus get $187 - 61 = 126$.

Comment. The result seems to have been known in China about 2000 years ago, and, possibly for that reason, it is usually referred to as the

Chinese remainder theorem (CRT), or, to be more accurate, as the first case of that theorem. The general case deals with more than two simultaneous congruences.

The conclusion of the general CRT (under a suitable assumption) is that the simultaneous congruences

$$x \equiv a_i \bmod m_i, \quad i = 1, \ldots, k$$

have a solution. The expression $my + nz$ plays a role in the case $k = 2$. In the present notation that looks like $m_1 y_1 + m_2 y_2$, and hence it looks as if the right thing to consider in the general case is

$$m_1 y_1 + \cdots + m_k y_k,$$

but it turns out that that is not helpful. The right trick is to set

$$m = m_1 \cdot m_2 \cdots m_k, \quad m_i' = \frac{m}{m_i},$$

and try to solve the special congruences

$$m_i' y_i \equiv 1 \bmod m_i, \quad i = 1, \ldots, k.$$

Once that's done, the desired solution becomes

$$x = \sum_{j=1}^{k} a_j m_j' y_j.$$

Indeed: $x \equiv a_i \bmod m_i$, and $x \equiv 0 \bmod m_j$ when $j \neq i$. The trouble is that in order to be able to solve the special congruences some assumptions are needed; the right thing to assume is that for each i the greatest common divisor of the numbers m_i' and m_i is equal to 1. That's a complicated way of saying that any two of the m_i's are relatively prime, and that is exactly the right assumption for the general CRT.

Details of the proof are not of enough interest to enter into them here, but it will be convenient to use the result in the next solution.

Solution 4 D.

4 D

The problem of finding two consecutive numbers neither of which is square-free is trivial; examples are 8, 9, and 24, 25, and 27, 28, and 44, 45. Is there, however, a method, a systematic way of producing such pairs?

Yes, there is, via the Chinese remainder theorem: apply that theorem to the (relatively prime) moduli 4 and 9 with the prescribed residues 1 and 2. That is: apply the Chinese remainder theorem to produce a number x that is congruent to 1 modulo 4 and at the same time congruent to 2 modulo 9. The method of the preceding solution is to find numbers y and z such that $9 \cdot y + 4 \cdot z = 1$; the simplest y and z that jump to the eye are $y = 1$ and $z = -2$. If the method is followed through with those numbers (a procedure that is warmly recommended—it will make the reasoning come alive), the result is $x = -7$; since in the present context positive numbers are wanted, that should be replaced by 29 $(= -7 + 4 \cdot 9)$. Check: $29 \equiv 1 \bmod 4$, and $29 \equiv 2 \bmod 9$. Equivalently: 27 $(= 29 - 2)$ is divisible by 9, and 28 $(= 29 - 1)$ is divisible by 4—here, indeed, are two consecutive numbers neither of which is square-free.

The significance of 4 and 9 resides in that they are squares and that they are relatively prime. (The latter assertion is obvious, of course, but the reason for it is worth stating explicitly; the reason is that 2 and 3 are distinct primes.) Once reasoning in this low case is understood, the result for the general case becomes easy to prove. The general result is that there exist arbitrarily long sequences of consecutive numbers none of which is square-free. To prove it for the length 6, as a typical example that may induce more confidence than the length 2, apply the Chinese remainder theorem to the moduli 4, 9, 25, 49, 121, and 169, to find a number x that is congruent, respectively, to 1, 2, 3, 4, 5, and 6. Write $n = x - 7$, and note that $n + 1$, $n + 2$, $n + 3$, $n + 4$, $n + 5$, and $n + 6$ are divisible by 169, 121, 49, 25, 9, and 4 respectively, and hence that none of them is square-free.

4 E Solution 4 E.

Some abbreviations come in handy; write n for 4444^{4444} and write S for "sum of the digits". What is wanted is $S(S(S(n)))$, or, more simply put, $S^3(n)$.

A very crude estimate, but one that is easy to calculate with, is that the base of the exponential expression 4444^{4444}, that is 4444, is less than or equal to 10,000. A somewhat less crude estimate is that the exponent is less than or equal to 5000. Consequence: $n \leqq 10{,}000^{5000}$. The latter number is 1 followed by 4×5000 zeroes; it follows that

$$S(n) \leqq 9 \times 20{,}000.$$

Consequence: $S(S(n))$ is less than or equal to the sum of the digits of each one of the positive integers below 180,000, and therefore

$$S(S(n)) \leqq S(99{,}999) = 45.$$

One more step of the same kind:

$$S(S(S(n))) \leqq S(39) = 12.$$

Now look at the situation modulo 9. Since

$$4444 \equiv 7 \bmod 9,$$

it follows that

$$4444^2 \equiv 49 \equiv 4 \bmod 9,$$

and

$$4444^3 \equiv 7 \times 4 = 28 \equiv 1 \bmod 9.$$

Think of 4444^{4444} written out as the product of 4444 factors each equal to 4444, and clump those factors in groups of 3 as far as possible. Since the number of factors is 4444, and since $4444 \equiv 1 \bmod 3$, the clumping will not be perfect; one factor of 4444 will stick out at the end. According to the last displayed congruence, the partial product contributed by each clump is congruent to 1 modulo 9, and, therefore, the whole product is congruent modulo 9 to the last factor that sticks out, which is congruent to 7 modulo 9.

What is known about $S^3(n)$ at this stage? Two things: one is that $S^3(n) \leqq 12$, and the other is that $S^3(n) \equiv 7 \bmod 9$. Those two pieces of information settle everything: there is only one positive integer that is below 12 and is congruent to 7 modulo 9, namely 7. The answer is 7.

Solution 4 F.

4 F

Contemplation of the prime factorization of positive integers shows that every number can be written in the form $h \cdot k$, where h is a power of 2 and k is odd. What happens if each of the numbers in a possible set with the non-divisibility property in question is written in that form? Could it happen that k, the odd factor, is the same for two different numbers in the set? No, it could not. Reason: if two numbers have the same odd factor, then they must be of the forms $h \cdot k$ and $h' \cdot k$, where h and h' are powers of 2. Since one of h and h' must always divide the other, it follows that one of the products $h \cdot k$ and $h' \cdot k$ must divide the other—which cannot be. Consequence: the odd factors of the numbers in any non-divisibility set (chosen from the integers between 1 and 100) must all be different. Since there are only 50 possible odd factors (only 50 odd numbers between 1 and 100), no set of the desired kind can have more than 50 elements. Equivalently: every set of 51 or more numbers between 1 and 100, must contain two of which one divides the other.

Comment. The generalization of the reasoning to numbers other than 100 is obvious. The resulting assertion is that every set of $n+1$ numbers chosen from among the first $2n$ positive integers contains two of which one divides the other.

Is that result the best possible one along these lines? That is: is it true, for every n, that it is possible to choose n numbers from among the first $2n$ so that none of them divides another? Yes—obviously; just choose the numbers $n+1, n+2, \ldots, 2n$.

There is a similar but considerably more trivial problem: does every set of $n+1$ numbers from among the first $2n$ contain two that are relatively prime? The answer is yes: every set of $n+1$ numbers from among the first $2n$ must contain at least two neighbors (that is, numbers such as k and $k+1$), and two neighbors are always relatively prime. Is that result best possible? Sure: the even numbers form a set of n no two of which are relatively prime. Are there any other examples?

Paul Erdős, the celebrated number theorist, used to enjoy looking for and finding mathematical prodigies, and he often used this problem as a preliminary test; to deserve to be called a prodigy, you must solve it instantaneously.

4 G Solution 4 G.

Denote the answer by e_n, (so that $e_1 = 2$, $e_2 = 4$, $e_3 = 6$), and ask how it changes as n changes. Each sequence of the kind described becomes another one if every 0 in it is replaced by 1, and vice versa. In other words, the sequences of the kind described come in pairs; the simplest way to distinguish between two elements of one pair is by looking to see if the first term is 0 or 1. Consequence: the unknown number e_n is twice the number of those among the sequences described that begin with, say, 0. Associate with each such sequence the sequence of 1's and 2's that counts the lengths of the blocks of equal digits, so that, for instance, the sequence associated with

$$0110010010$$

is

$$1221211.$$

The lengths of the blocks in a sequence add up to the length of the sequence, and, therefore, the sum of the sequence of 1's and 2's so associated with a sequence of 0's and 1's of length n is n. If the sequence of 1's and 2's ends in 1, then the sequence obtained from it by erasing its last term has sum $n-1$; if it ends in 2, then after the erasure it has sum $n-2$. This implies that the numbers $\frac{1}{2}e_n$ have the property that each one (after the second) is the sum of its two predecessors. (Is that clear? If the erased last count is 1, then the original sequence of length n is obtained from a sequence of length $n-1$ by tacking on 0 or 1, depending on whether the sequence of length $n-1$ ended in 1 or 0; if the erased last count is 2, then the original sequence of length n is obtained from a sequence of length $n-2$ by tacking on 00 or 11, depending on whether the sequence of length $n-2$ ended in 1 or 0—and those are the only ways of getting a sequence of length n that has no 3-blocks.) In other words,

$$\frac{1}{2}e_n = \frac{1}{2}e_{n-1} + \frac{1}{2}e_{n-2}.$$

It's more comfortable to avoid fractions; put $f_n = \frac{1}{2}e_n$, and rewrite the equation as

$$f_n = f_{n-1} + f_{n-2}.$$

This form of the equation is called the Fibonacci recursion equation and the numbers satisfying it (beginning with $f_0 = f_1 = 1$) are the well-known Fibonacci numbers. The sequence is fun to contemplate even if you don't do anything with it; the first few terms are

$$0,\ 1,\ 1,\ 2,\ 3,\ 5,\ 8,\ 13,\ 21,\ 34.$$

In this language the answer to the question is $2f_{n+1}$.

Comment. This is really a combinatorial problem, but it appears here, under number theory, because of its Fibonacci connection.

4 H

Solution 4 H.

If $N = 1, 2, 3$, or 4, the problem is trivial; there is no real loss in assuming that $N \geqq 5$.

Can any of the parts in the maximal partition of N be equal to 1? No, that's silly; erasing that 1 and increasing any other part by 1 will maintain the sum and increase the product.

Can any of the parts in the maximal partition of N be 4 or more? If there is a part $p \geqq 4$, replace it by $(p-2)+2$. The effect of that replacement is to replace a factor p of the product by $2p - 4$. The assumption $p \geqq 4$ implies that $2p - 4 \geqq p$, which means that the replacement does not decrease the product, and might increase it. Conclusion: to get a maximal partition, only the parts 2 and 3 need to be used.

Can there be more than two parts of size 2 in the maximal partition of N? No; the replacement of 2+2+2 by 3+3, has the effect of replacing a factor 8 of the product by a factor 9.

Conclusion: use as many 3's as possible, and then finish by using no 2's, or just one, or two of them, depending on whether $N \equiv 0$ or 2 or 1 modulo 3.

The preceding paragraphs can be summarized as follows.

If N is divisible by 3, then the maximum product is obtained from partitioning N into $\frac{N}{3}$ parts all equal to 3; the value of the maximum product in that case is $3^{N/3}$.

If N is congruent to 2 modulo 3, then partition N into $\frac{N-2}{3}$ parts equal to 3 plus one part equal to 2; the value of the maximum product in that case is $2 \cdot 3^{(N-2)/3}$.

If, finally, N is congruent to 1 modulo 3, then partition N into $\frac{N-4}{3}$ parts equal to 3 plus two parts equal to 2; the value of the maximum product in that case is $4 \cdot 3^{(N-4)/3}$.

Comment. For the special case $N = 10$, these considerations yield the partition $3+3+2+2$, with product 36—which is what the preliminary experimentation indicated. For $N = 11$, the best partition is 3+3+3+2, with product 54.

Solution 4 I.

4 I

The answer is yes. Since the powers of 10 are easy enough to calculate, the main piece of information that is needed concerns 5^{10}, namely that $5^{10} \equiv 1 \bmod 11$. That's easy: since

$$5^2 = 25 \equiv 3 \bmod 11,$$

it follows that

$$5^4 \equiv 3^2 = 9 \equiv -2 \bmod 11,$$

and hence that

$$5^5 = 5^4 \times 5 \equiv (-2) \times 5 = -10 \equiv 1 \bmod 11.$$

Since $10 \equiv -1 \bmod 11$, and since every power of 5 is odd, it follows that

$$10^{5^{10^{5^{10}}}} \equiv -1 \bmod 11.$$

Now use the fact about 5^{10}; since $10^{5^{10^5}}$ is a multiple of 10, it follows that

$$5^{10^{5^{10^5}}} \equiv 1 \bmod 11.$$

Comment. The assertion $5^{10} \equiv 1 \bmod 11$ is a special case of the so-called "little Fermat theorem", which says that if p is a prime and $x \not\equiv 0 \bmod p$, then $x^{p-1} \equiv 1 \bmod p$. A slick proof of that, in turn, is the observation that the non-zero integers modulo p form a group of order $p-1$ with respect to multiplication modulo p.

Solution 4 J.

4 J

If a power of 3 is less than another power of 3, then the smaller one remains smaller even after it's multiplied by 2. Indeed: if $3^x < 3^y$, then $x < y$, so that $x + 1 \leqq y$, and therefore

$$2 \cdot 3^x < 3 \cdot 3^x = 3^{x+1} \leqq 3^y.$$

Suppose now that

$$9^{(m)} < 3^{(n)}$$

for some positive integers m and n. Since $9^{(m)}$ is a power of 3,

$$9^{(m)} = 9^{9^{(m-1)}} = (3^2)^{9^{(m-1)}} = 3^{(2 \cdot 9^{(m-1)})},$$

the first observation applies, and it follows that

$$2 \cdot 9^{(m)} < 3^{(n)}.$$

Consequence:

$$9^{(m+1)} = 3^{(2 \cdot 9^{(m)})} < 3^{3^{(n)}} = 3^{(n+1)}.$$

What was proved so far is that once an inequality such as $9^{(m)} < 3^{(n)}$ happens to be true, then it remains true when both the superior indices are augmented by 1. The proof is over with a wave of the magic induction wand; just start from $9^{(1)} < 3^{(2)}$, and infer that

$$9^{(m)} < 3^{(m+1)}$$

for every positive integer m. That result contains, in particular, the special case

$$9^{(100)} < 3^{(101)},$$

which surely cannot be improved.

4 K Solution 4 K.

The answer is certainly greater than 2. Reason: to any two distinct points there corresponds at least one point whose distance from each of the given ones is rational. Proof: draw circles centered at the given points with rational radii; if the circles are not chosen too carelessly, they will intersect. (Choose the radii to be more than half the given distance but less than the whole.)

Three punches are enough, and once that result is suspected it is difficult to avoid proving it. Almost any three points in the plane have the property that every point is at irrational distance from at least one of them. The first three you might think of trying are $(0, 0)$, $(1, 0)$, and $(\sqrt{2}, 0)$—and they work just fine.

What does it mean to say that they work? It means that for every point (x, y) in the plane at least one of the distances

$$\sqrt{x^2 + y^2}, \qquad \sqrt{(x-1)^2 + y^2}, \qquad \sqrt{(x-\sqrt{2})^2 + y^2}.$$

is irrational. In fact slightly more is true: at least one of the squares of those distances is irrational. To see that, reason as follows. If either

$$x^2 + y^2 \qquad \text{or} \qquad (x-1)^2 + y^2$$

is irrational, we're done. If they are both rational, then they form a conspiracy to prove that the middle term of $(x - 1)^2$ is rational, which implies, of course, that x is rational. The conspiracy continues: the known rationality of both $x^2 + y^2$ and x, and the occurrence of $\sqrt{2}$ in the middle term of $(x - \sqrt{2})^2 + y^2$ implies that the latter cannot be rational, and that's that.

This is a proof, but it's not the proof that most mathematicians would prefer. It requires no thought and it requires some computation, admittedly tiny, but still computation. The "correct" proof requires thought but absolutely no computation; it goes like this.

Punch twice, at distinct centers. Since each punch leaves countably many circles, the two punches leave their intersections, a countable set. Consider all circles centered at points of that set, with rational radii; their intersections with an arbitrary line form a countable set. A point of that line not in that countable set is at an irrational distance from all remaining points; apply the punch there.

Chapter 5. Geometry

Solution 5 A.

5 A

It is tempting to say that the answer is 7, because $5 + 4 - 2 = 7$. The idea is that the pyramid has five faces and the tetrahedron has four, and the gluing removes two faces (one from the pyramid and one from the tetrahedron). That reasoning could be correct, but before it is accepted it is necessary to examine whether the faces form any conspiracies.

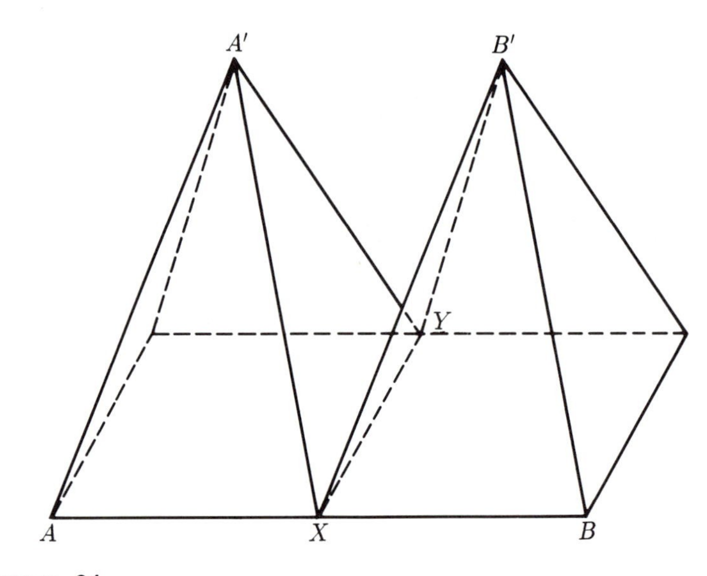

FIGURE 34

The inspired way to make the examination is to put two copies of the pyramid next to each other so that the their square bases sit on the same plane and share an edge, as in the figure. What can be said about the gap between them?

Observe, to begin with, that the front triangles $AA'X$ and $BB'X$ are in the same plane, and then look at the triangle $A'XB'$ between them in that plane. The altitude of $AA'X$ is the length of the perpendicular (in that plane) from A' to the segment AX, and the altitude of $A'XB'$ is the length of the perpendicular (in that plane) from X to segment $A'B'$. The segments AX and $A'B'$ are parallel segments in the same plane, and both of the altitudes just described are equal to the distance between them. Consequence: the triangle $A'XB'$, which is obviously isosceles, has the same altitude as the equilateral triangle $AA'X$. Since the two sides of $A'XB'$ that are known to be of equal length have the same length as the sides of $AA'X$, it follows that $A'XB'$ must be congruent to $AA'X$, and, therefore, it must be equilateral. That's not obvious to begin with, but once it becomes known everything is settled.

The main piece of information that the equilateral character of $A'XB'$ gives is that the segment $A'B'$ has the same length as the sides of all the triangles that enter the story. Consequence: the gap between the two copies of the pyramid P is exactly the size of the regular tetrahedron T. In other words, if a face of T is glued onto the face $A'XY$ of P, the gap between the two pyramids is exactly filled, and, more importantly, the face $A'XB'$ is in the same plane as $AA'X$. The right-hand copy of the pyramid can now be thrown away, and what remains shows that the faces did form coplanarity conspiracies. What remains is, in fact, the glued surface with the same base as P, having one rhombus face in front and one in back, and one triangle face at left and one at right—a total of five faces.

Comment. The story goes that this problem appeared once on one of those "are-you-smart-enough-to-get-into-college" tests, like the SAT, and the correct answer, here reported was given zero credit—the designers of the test thought that the right answer was 7. The victim protested, and much to the surprise of the graders, won his protest.

Solution 5 B.

5 B

Yes, a join closed finite set in the plane must be a subset of a line. One way to arrange the proof is by contradiction: assume that a non-linear join closed finite set exists and run into trouble.

If F is a join closed finite set in the plane, consider all the lines that go through two points of F and consider for each point of F the distance from that point to any of those lines that do not contain it. The set of all distances so obtained is finite; let p be a point and L a line such that $d(p, L)$, the distance from p to L, is the minimum value of those distances. Since F was assumed to be non-linear, $d(p, L)$ is positive. What trouble can that lead to?

Where does the foot of the perpendicular from p to L hit L?

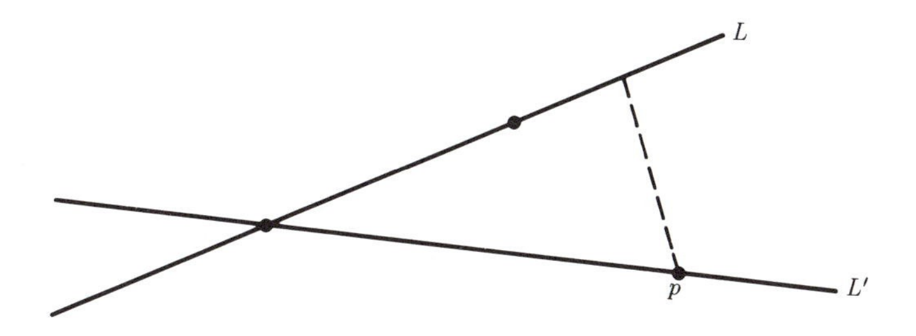

FIGURE 35

1. If both the original points that caused L to be in the competition at all are on the same side of the foot of the perpendicular, then consider the line L' that joins the farther one to p, and consider the distance from the nearer one to L' (see Figure 35). Since that distance is smaller than $d(p, L)$, the minimum, there is trouble.
2. If the foot of the perpendicular is between the original points that put L into the competition, then, by the join closed assumption, there is at least one other point of F on L, and it must be on the same side of the foot of the perpendicular as one of the original points (see Figure 36). The situation has returned to case (1)—trouble again.

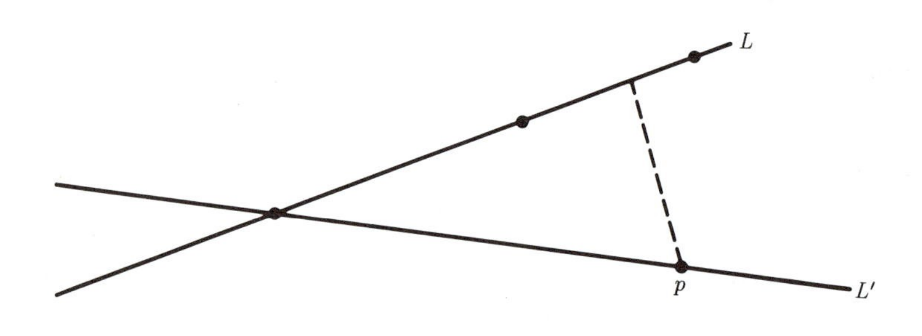

FIGURE 36

Conclusion: the assumption that $d(p, L) > 0$ is untenable; the set F must be a subset of a line after all, and the proof is complete.

Comment. The problem was originally proposed in 1893 by the great English geometer J. J. Sylvester, who could not solve it. He thought, in fact, that it was very hard, and I have heard it reported that he thought Fermat's last theorem (Fermat's last conjecture?) would be solved first. Well, Fermat is still unsolved, but the brilliant argument about distances stands before us and shows that the answer to Sylvester's question is yes. It did, to be sure, take over fifty years to find the answer.

The argument was discovered by Tibor Gallai, and several modern professional geometers don't like it—they disapprove of it. The reason for their negative judgment is that the argument depends on distances, whereas the question uses only the language of projective or affine geometry. “Pure” proofs have been discovered since Gallai’s breakthrough, but they are harder, deeper, more complicated.

Solution 5 C.

5 C

If all three vertices of a triangle inside a square of area 1 are vertices of the square, the area of the triangle is $\frac{1}{2}$. If at least one vertex, say C, of the triangle, is not a vertex of the square, consider a line through C parallel to the sides of the square perpendicular to the side that C is on; that line will separate the triangle into two smaller triangles.

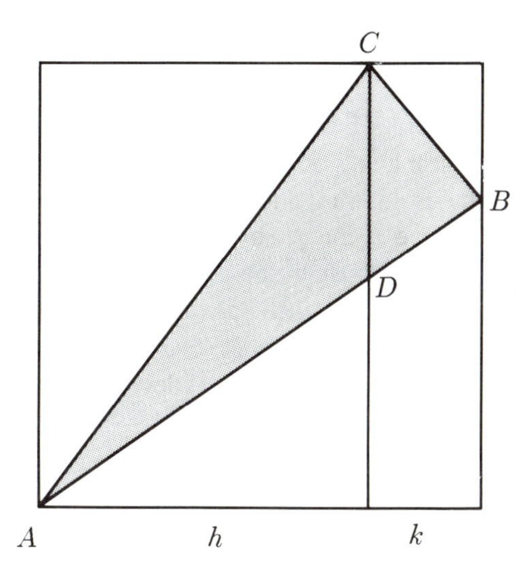

FIGURE 37

The figure now tells the story:

$$2 \text{ area } ABC = CDh + CDk = CD \cdot (h + k) \leqq (h + k)^2 = 1.$$

Equality can be attained, of course; that happens if and only if AB is a side of the square. Conclusion: the answer is $\frac{1}{2}$.

Comment. Could the question have been asked the other way around: for each triangle of area 1, what’s the largest area that a square inside it can have? No, a pleasant answer cannot really be expected. Think, for instance, of a triangle with altitude equal to one millimeter and base equal to two million millimeters (= two kilometers). Its area is one square meter, but it cannot contain any square as big as one square millimeter.

Here is a related question with a differently placed quantifier: what is the largest area that a square inside *some* triangle of area 1 can have? Think for instance of an isosceles right triangle of area 1 (sides $\sqrt{2}$ and $\sqrt{2}$, and hypotenuse 2), and nestle a square in the right angle with one vertex on the hypotenuse—it has area $\frac{1}{2}$. Can that area be made larger? The answer is known to be no; the proof is not easy.

5 D

Solution 5 D.

The problem is to find a minimum, and, as such, it is a more or less standard calculus problem. Looking at it geometrically, however, yields a point of view that is different and refreshing. Here is how.

Draw the positive x and y axes and the given point, and reflect the picture so obtained through the given point. A line with negative slope through the given point (such lines are the only ones that enter the competition) determines two triangles: one with the original positive x and y axes and another one with their reflections. The union of those two triangles covers the rectangle bounded by the positive x and y axes

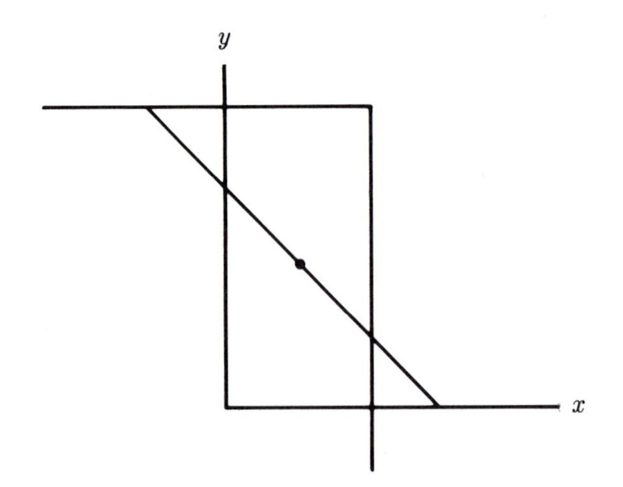

FIGURE 38

and their reflections, and, therefore, has area greater than or equal to the area of the rectangle; equality occurs if and only if the line (the diagonal?) passes through the originally given point. Conclusion: the optimal line is the diagonal of the rectangle.

Solution 5 E.

5 E

There is a road system that is shorter than the two diagonals, namely one of the form seen in Figure 39.

Once that idea presents itself, a little thought and a little calculus can help. If $x = \frac{1}{2}$, the length of the road is $\frac{1}{2} + \sqrt{5} = 2.736\ldots$. The minimum length among roads of this form (via calculus) occurs when $x = 1 - \frac{\sqrt{3}}{3}$; its length is $1 + \sqrt{3} = 2.732$.

Despite this satisfactory answer, an honest scholar should admit that the original question is not completely answered. To prove that the road systems of the form indicated above is shorter than any other takes proof—it takes thinking along the lines of the calculus of variations.

Comment. What is the answer for different numbers of houses distributed differently: for instance for an equilateral triangle, for an isosceles right triangle, for an arbitrary triangle, for more general (regular?) polygons?

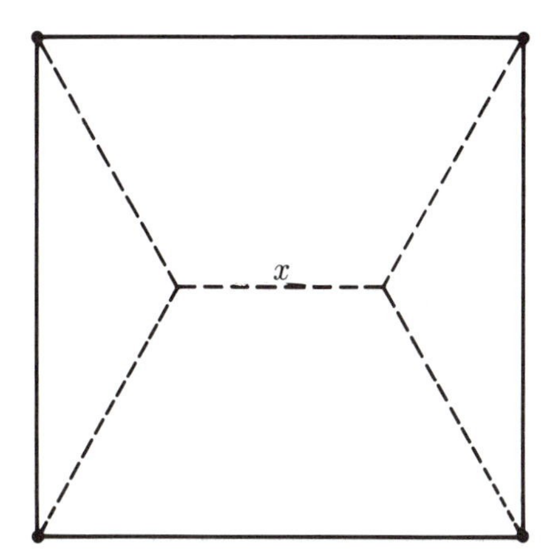

FIGURE 39

5 F Solution 5 F.

Pave enough of the plane with small squares to be sure that both the given points are in one of the squares. The word "small" here should be interpreted to mean small enough to make Euclidean constructions with a short straightedge and a rusty compass possible—an eighth of an inch is surely small enough.

Select a point within each small square, and then construct a polygonal path that has the selected points as vertices and that joins the two originally given points. The meaning of those instructions is clear, isn't it? They mean, choose one of the originally given points to start at, select the small square containing that point, join the point to the point in an adjacent small square (with the straightedge), then join that to the point in another adjacent small square, and keep going till the second originally given point is reached. All that is needed to make the process possible and legal is the short straightedge, the rusty compass, and mathematical induction.

Look at the polygonal path so constructed, count the number of segments it consists of, and then reduce the entire picture by a factor so much greater than their number that the entire path can be reproduced (homothety) inside the small square that contains the starting point. The resulting miniaturized photograph includes, in particular, the image of the ending point, and, consequently, that image and the starting point can be joined by a line segment (using the short straightedge). That's it—that's the next to the last step to victory—that line segment points in the right direction. All that remains to be done is to keep using the straightedge over and over again to extend that straight edge—after a while the extension so obtained will pass through the ending point.

5 G Solution 5 G.

Consider all possible straight lines that have on them two points of the 3000; there are a lot of them, but still only a finite number. It is therefore possible to find a straight line that is not parallel to any of them. Let L be such a straight line, far away from the scene of the action, meaning a straight line such that every one of the 3000 points is on the same side of it.

Once such a line has been found, the work is over. What now has to be done is to move the line L parallel to itself toward the 3000 points till the first time it hits one of them, keep moving till the second time, and keep moving till the third time—and then stop. The three points so picked up determine a triangle, and that triangle is the first of an inductively defined set. Just keep moving L parallel to itself—as it moves it picks up points one at a time, and every time it picks up three new points, they determine a triangle disjoint from all the triangles found so far.

The answer to the question is yes: if no three among $3n$ points are collinear, then there exist n disjoint triangles that have those $3n$ points for their vertices.

Solution 5 H.

5 H

The answer can be determined by calculus methods, but there is a better way.

There exists a linear transformation that maps the ellipse onto a circle. Linear transformations do not necessarily preserve areas, but they do preserve their order—if one rectangle is smaller than another before the transformation, it will remain smaller after. Consequence: a linear transformation of the kind being considered will map the largest inscribed rectangle in the ellipse onto the largest inscribed parallelogram in the circle, which means that it will map that maximal rectangle onto the inscribed square. Since linear transformations also preserve the ratio of areas, it follows that the unknown ratio for the rectangle is the same as the easily calculable one for the circle. Conclusion: the ratio is

$$\frac{\pi}{\left(\frac{2\sqrt{2}}{2}\right)^2} = \frac{\pi}{2}.$$

The datum about major and minor axes is a red herring.

Comment. This solution indicates, once more, the unity of mathematics—something that looks superficially like a messy calculus problem can, in fact, be effectively and neatly solved by linear algebra.

Solution 5 I.

5 I

For an isosceles right triangle the picture suggested that the most efficient bisecting curve is the perpendicular from the right angle to the hypotenuse, whose length is the altitude. (By "the" altitude of an isosceles triangle, I mean the one between the two equal sides.) For some isosceles triangles the altitude is equal to the length of the bisecting segment parallel to the base (compute the altitude of the triangle with sides $(\sqrt{3}, \sqrt{3}, 2)$). For an equilateral triangle with side length 1 a little calculation shows that the altitude is $\frac{\sqrt{3}}{2} = .866\ldots$, whereas the length of the bisecting segment parallel to one of the sides is $\frac{\sqrt{2}}{2} = .707\ldots$. The altitude might have seemed like a reasonable first guess, but more thought is obviously needed. A little more meditation seems to yield the conclusion that the most efficient way to bisect an equilateral triangle is to connect two sides—the question is how. There is a trick that reveals the answer.

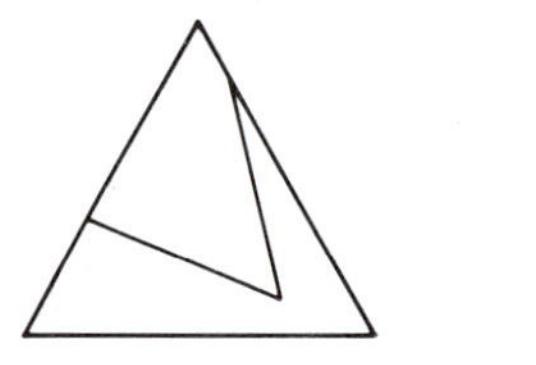

FIGURE 40

Look at a curve that connects two sides of an equilateral triangle and bisects its area (Figure 40), and then reflect the triangle, curve and all, through one of those two sides (Figure 41). Reflect the new

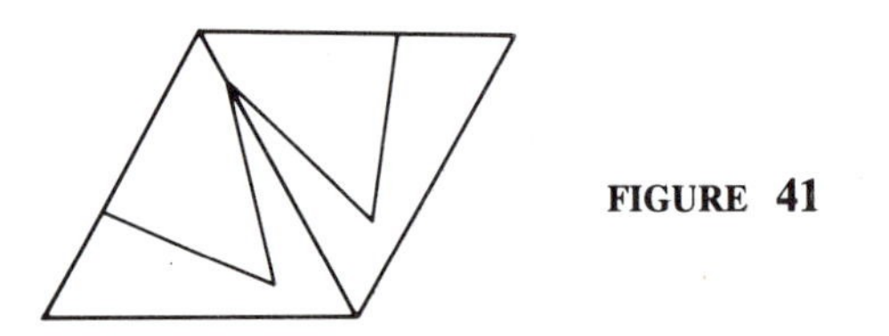

FIGURE 41

triangle, curve and all, about the image of the other side involved (Figure 42), and keep reflecting till the result is a regular hexagon inside which there is a closed curve that surrounds the center (Figure 43).

How much of the area of the hexagon is inside that closed curve? The answer is obvious, isn't it? Since the original curve bisected the area of the triangle it was in, its successive images bisect the triangles they

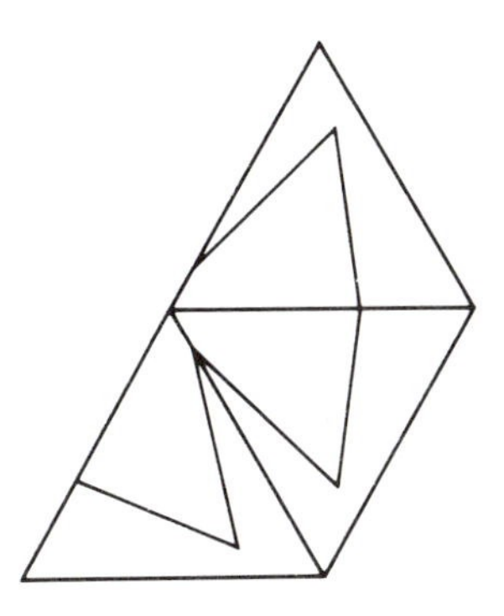

FIGURE 42

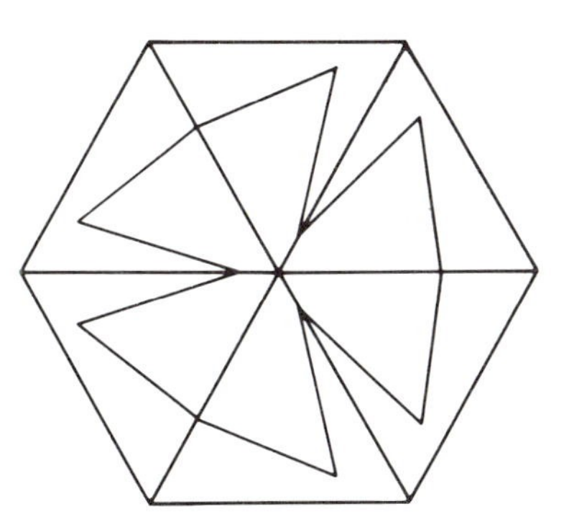

FIGURE 43

find themselves in, and, therefore, the entire closed curve bisects the hexagon: it contains exactly a half of the area of the hexagon.

How long is that closed curve? Answer: six times as long as the original connecting curve. How can that answer be made as small as possible? In other words: given a regular hexagon, what's the shortest closed curve inside it that contains exactly half the area? At this point the word "half" is pointing us in the wrong direction—let us ignore it. We started with a given equilateral triangle, which has a definite area; what we are now looking for is the shortest closed curve (in a hexagon) that surrounds exactly three times (half of six times) that area. What is the shortest closed curve that surrounds any given area? The answer to that has been officially known for over two thousand years and is intuitively obvious to everybody: the answer is the circle with the prescribed area. The most efficient way to bisect an equilateral triangle is with the

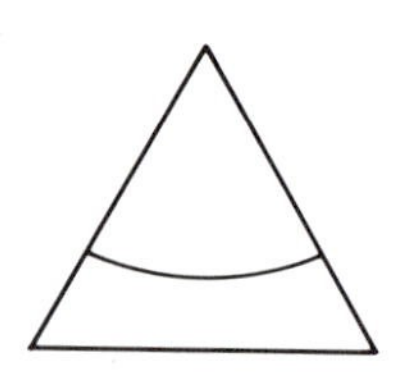

FIGURE 44

arc of a circle with center at one of the vertices (Figure 44). Isn't that (a) pretty, (b) surprising, and (c) obvious (now that we see it)?

Comment. To calculate the length of the circular arc is not difficult, but slightly messy; it comes out to be

$$\frac{\pi}{3}\sqrt{\frac{3\sqrt{3}}{4\pi}} = .673\ldots,$$

which is a lot shorter, as such things go, than the .866... of the altitude and the .707... of the parallel bisector.

5 J

Solution 5 J.

The answer is yes: two triangles of the same area are always Cavalieri congruent. What the proof must try to accomplish is to position two given triangles of the same area so that all their cross sections have the same lengths in some direction. There is a sense, however, in which it is enough to win the game in one direction only—provided that the score is somewhat better than the bare minimum needed for a win.

Suppose, in fact, that T $(= (ABC))$ and T' $(= (A'B'C')')$ are triangles of area 1 (a normalization is surely harmless), and suppose that for some points X and X' on the sides BC and $B'C'$ (opposite to A and A') it is true that the lengths of AX and $A'X'$ are equal, and, moreover, that the areas of the subtriangles ABX and $A'B'X'$ are equal. It follows that the altitude from B to the line of AX is equal to the altitude from B' to $A'X'$ (Figure 45). Assertion: it follows automatically that the lengths of all the cross sections parallel to AX (and $A'X'$) are equal. The reason is similarity. Think of AX as sliding toward B, remaining parallel to its original position (and with end points remaining on the sides AB and BC). If A^+X^+ is one of the possible positions of AX, then the triangles

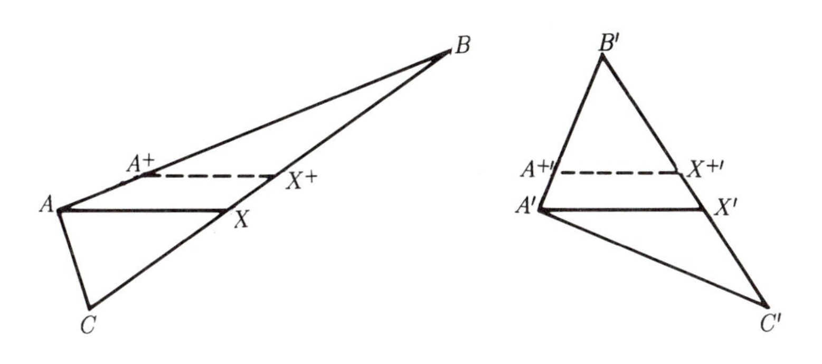

FIGURE 45

ABX and A^+BX^+ are similar (angles are equal). If $A'X'$ is slid toward B' the same way by the same amount, then $A'B'X'$ and $A^{+\prime}B'X^{+\prime}$ are similar to each other. Since the altitudes of A^+BX^+ and $A^{+\prime}B'X^{+\prime}$ are equal, it follows that their areas are equal, and therefore their bases A^+X^+ and $A^{+\prime}X^{+\prime}$ have equal lengths. The same argument works in the other direction, toward C and C', and the conclusion is that T and T' are Cavalieri congruent, as promised.

In view of the preceding discussion, the problem reduces to a study of the relation between certain lengths and areas, namely the lengths of segments from the vertices of a triangle to the opposite sides and the areas that such segments cut off. A systematic way to study that relation is to "sweep" over a triangle T with segments from each vertex, one after another. Such a sweep covers the triangle T three times, so that the cumulative area goes from 0 to 3. Here is how the details look.

As a point X moves from B to C, the area of the triangle ABX

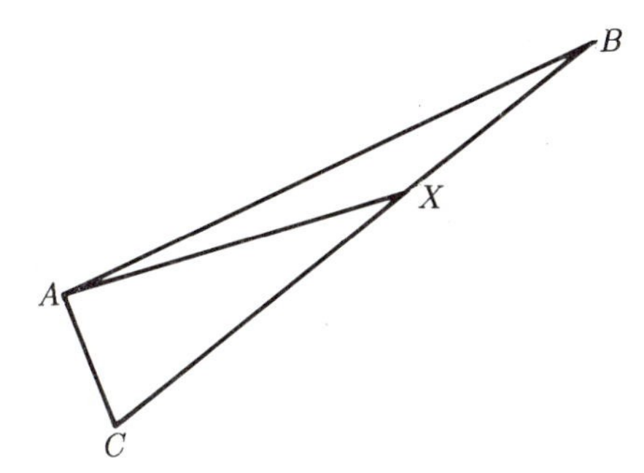

FIGURE 46

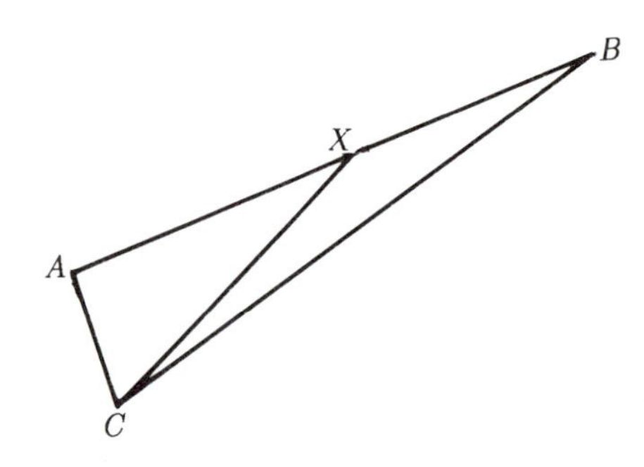

FIGURE 47

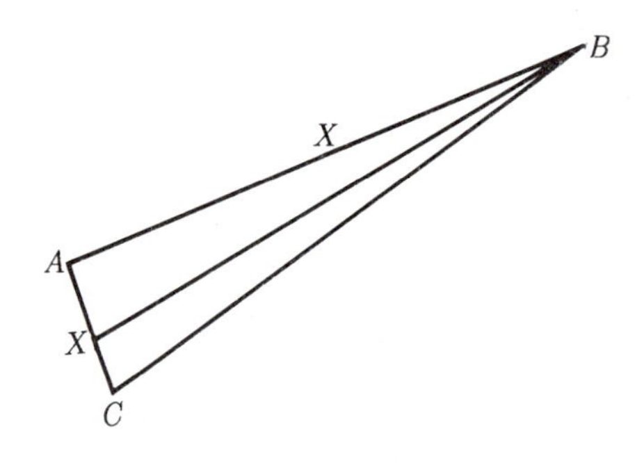

FIGURE 48

goes from 0 to 1. Then start using C as the pivot, and let X on AB move from A to B—the cumulative area, which is $1 +$ area (AXC), will then go from 1 to 2 (Figure 47). Finally, pivot from B, with X on AC (Figure 48) moving from C to A, as the cumulative area, which is $2+$ area (BCX), goes from 2 to 3. The lengths of the segments from the vertices to the moving point X vary, of course, as the area changes; the length function starts at $f(0) =$ length AB, goes to $f(1) =$ length AC, continues to $f(2) =$ length BC, and ends up where it started with $f(3)$ $=$ length AB. In other words, in better words, the length of a segment that cuts off a certain (cumulative) area is regarded as a function of that area; in this way the triangle T induces a (continuous) function on the interval $[0, 3]$. There is some freedom of choice in the construction—we can start at any vertex and go from either one of the sides that starts there to the other. Fortunately, for present purposes it doesn't matter how the choices are made—they are all equally useful.

Recall now that if a triangle is replaced by a similar one whose sides bear the ratio r to the corresponding sides of the given one, then the area of the new triangle bears the ratio r^2 to the area of the old. Our present view is in reverse: a "linear" measure (length) is regarded as a function of a "planar" measure (area), so that the length function under study looks (locally) like a piece of the graph of $y = \sqrt{x}$. For the particular triangle in the figures above, the length shrinks from AB to the altitude from A to BC, then rises to AC, continues rising to BC, and finally shrinks back to AB; the graph looks something like Figure 49.

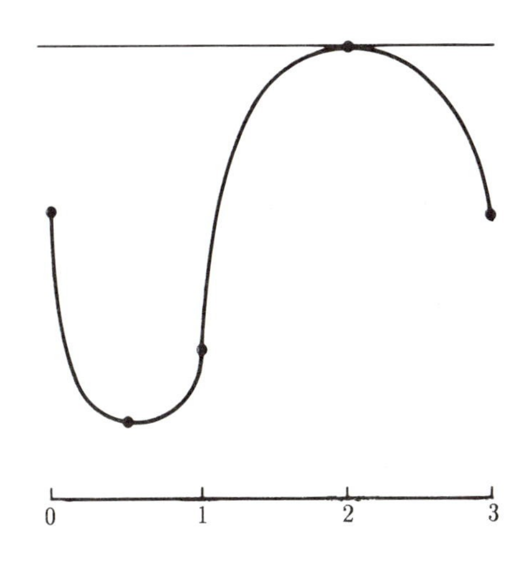

FIGURE 49

A continuous function on a compact interval attains its maximum and minimum values: what are those in the present case? Isn't the answer obvious? The maximum value of the length function is the length of the longest side of T, and the minimum value is the shortest altitude (the one perpendicular to the longest side). It is a useful corollary of this observation that the product $\max \cdot \min$ is equal to 2—twice the area of T.

All this was moderately hard work, but it's valuable—and now it's over. What conclusions does it lead to? Answer: two triangles T and T' of area 1 define two continuous functions f and f' on $[0, 3]$ each of which has the property that $\max \cdot \min = 2$. Consequence: the graphs of the functions f and f' have at least one point of intersection. That consequence is in a certain sense "obvious", but it's worth a careful look anyway. If it does *not* happen that $f(x) = f'(x)$ for some x in $[0, 3]$, then one of the functions f and f' is strictly greater than the other for all x; it is just a question of notation to assume that $f(x) > f'(x)$ for all x. From that it follows that there exists a positive number ε such that

$$f(x) \geqq f'(x) + \varepsilon \quad \text{for all } x.$$

This implies that $\max f \geqq f'(x) + \varepsilon$ for all x, and hence that

$$\max f \geqq \max f' + \varepsilon.$$

Since at the same time $f(x) \geqq \min f'$ for all x, which implies that

$$\min f \geqq \min f',$$

it follows that

$$2 = \max f \cdot \min f \geqq (\max f' + \varepsilon) \cdot \min f' = 2 + \varepsilon \min f'.$$

Since the length function f' can never take the value 0, so that $\min f'$ is strictly positive, a contradiction has arrived—the assumption that the graphs of f and f' do not intersect is untenable.

To say that $f(x_0) = f'(x_0)$ for some x_0 is the same as to say that there exists a cross section of the triangle T and a cross section of the triangle T' such that the two are of equal length and cut off equal areas. As the discussion at the beginning of this proof showed, that implies everything.

Comment. This is an astonishing result that seems to have gone unnoticed until its relatively recent discovery by Howard Eves. The proof here presented may appear verbose, and, indeed, proofs of the result can be given in many fewer words—but the result is subtle and, surely, the boredom that a few possibly unnecessary words induce is outweighed by the clarity they can achieve.

5 K

Solution 5 K.

The answer is yes, but it seems to be easier to work with the complement and to prove that the set of all those points in the plane that are *not* extreme points of a given closed convex set K is open. If a point E of the plane is *not* an extreme point of K, then either it is not a point of K

at all, in which case, clearly, it has a neighborhood that does not meet K (since K is closed), or it is a point of K, in which case it lies in the "interior" of some segment (Figure 50) joining two points A and B of K.

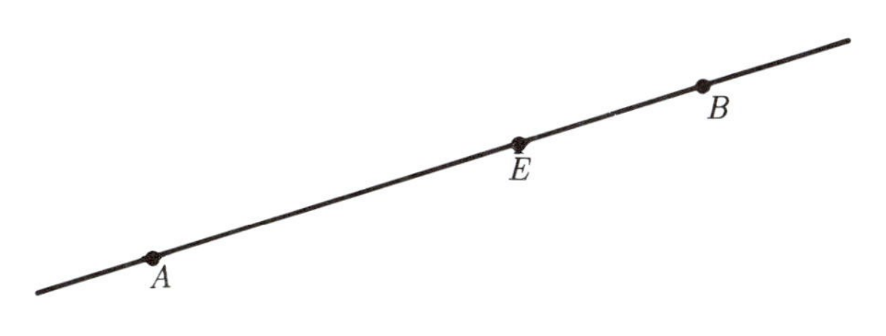

FIGURE 50

Consider a neighborhood of E (Figure 51) that contains neither A nor B. If there is such a neighborhood that contains no extreme points of K, then E is an interior point of the set of non-extreme points, and all

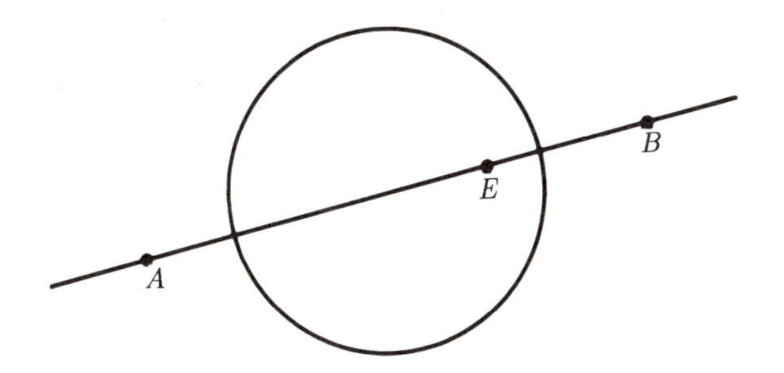

FIGURE 51

is well. If, however, the selected neighborhood does contain an extreme point P of K, then construct a neighborhood of E that contains none of A, P, B (Figure 52), and ask whether that new, smaller, neighborhood contains any extreme points of K. If the answer is no, then, as before, all is well. If, however, that new neighborhood contains an extreme point Q of K, then observe, in the first place, that Q is not inside the triangle ABP (because the interior points of that triangle are not extreme points), and then form (Figure 53) the quadrilateral $APBQ$ (interior and boundary). Since, by convexity, that entire quadrilateral is included

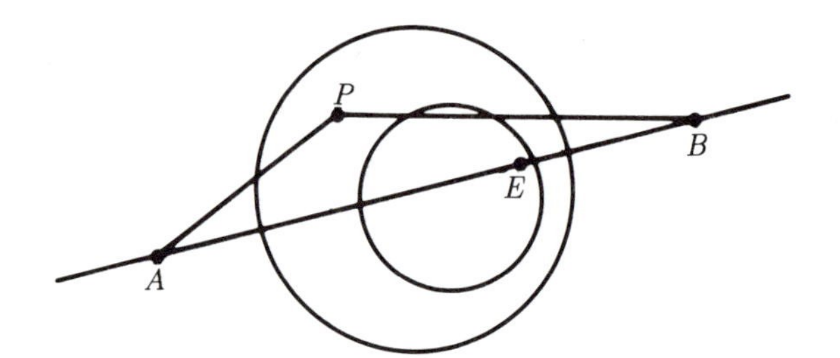

FIGURE 52

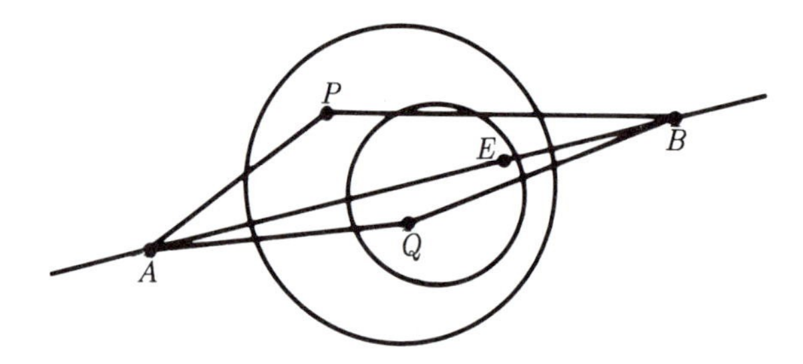

FIGURE 53

in K, its interior is a neighborhood of E that contains no extreme points of K, and the proof is complete.

5 L

Solution 5 L.

The answer is no—the set of extreme points of a compact set in three-dimensional space may fail to be closed.

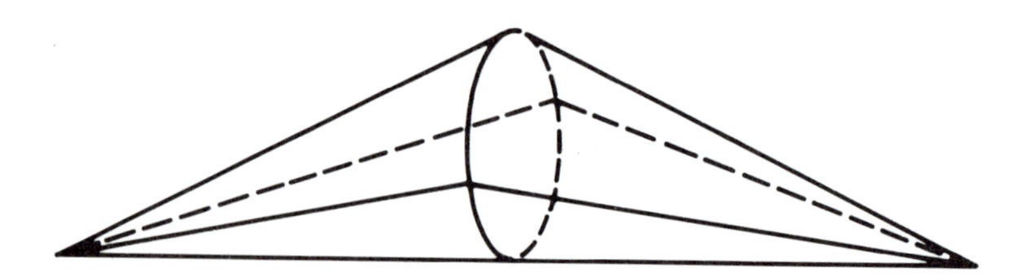

FIGURE 54

Once that answer is guessed, there is a natural way to go about trying to construct an example: consider a natural non-extreme point (in, say, an interval) and a natural extreme point (on, say, a circle), and put them together. Here is one concrete way to do that: put a closed interval on the plane of the page, put a disk (circle with interior) in a plane perpendicular to the page, tangent to that interval at its midpoint, and then join every point of the disk to every point of the interval. (Pertinent technical word: form the convex hull of the interval and the disk.) The result is a closed convex set. The points on the perimeter of the disk want to be extreme points, and all but one of them succeed: the one that fails is the midpoint of the original interval. That point of the closed convex set is a limit point of extreme points, but is not an extreme point itself.

Comment. There are many other examples. One of them is a cubical house with a cupola on top. Make sure that each point of the cupola, except of course at the edges that it shares with the cube, is "rounded", so that each such point is an extreme point; the points inside the edges of the cube are not.

Solution 5 M.

5 M

The answer is yes, and one way to prove it is by induction on the dimension n. When $n = 1$ all that is needed is to realize that a compact convex set in the line is a closed interval. In the general case the proof uses an

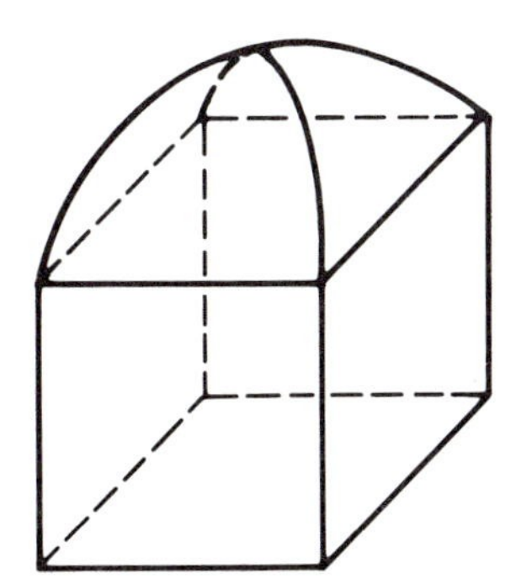

FIGURE 55

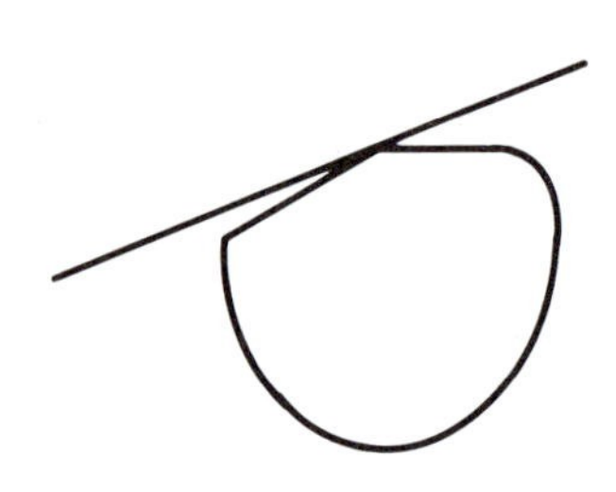

FIGURE 56

auxiliary concept, of geometric interest and importance in its own right, namely the concept of support.

A line of support of a convex set K in the plane is a line such that K is entirely on one side of it, and such that if the line is moved parallel to itself toward that side, however little, then K starts appearing on the hitherto empty side. A plane of support of a convex set in space is defined similarly, and the concept extends to higher-dimensional spaces (hyperplanes of support) with no trouble.

Suppose now that K is a compact convex set in $\mathbb{R}^n$, and consider two points, a and b say, in K. Form the ray $a \to b \to \infty$, and let c be the last point of K on that ray. Construct a supporting hyperplane to K at c, and intersect it with K; the result is a compact convex subset of K, in a space of one lower dimension. By induction, that set has an extreme

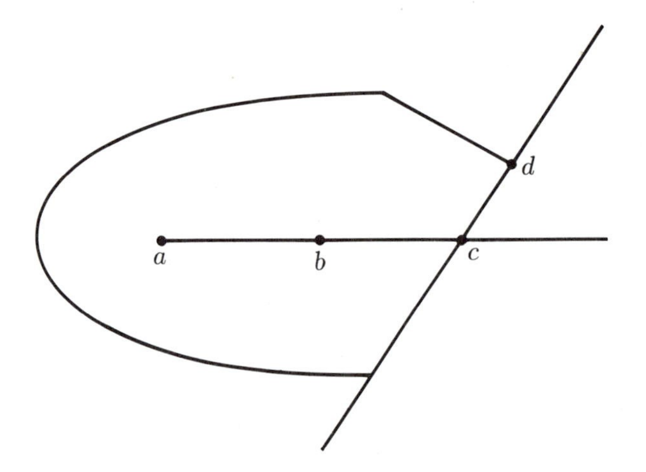

FIGURE 57

point, say d. Assertion: d must be an extreme point of K. Reason: if d is not an omittable point of K, then it is interior to some segment $[p, q]$ in K, and then both p and q must be on the hyperplane constructed in Figure 57, for otherwise d couldn't be; and if p and q are on that hyperplane, then d couldn't have been an extreme point of the lower-dimensional set.

Comment. The result is called the Krein–Milman theorem. The theorem has an infinite-dimensional version, and that, in fact, is the one that is most frequently applied in analysis—one of its applications is in the proof of existence theorems. (The conclusion is, after all, that certain things—extreme points—exist, and if the set in question is sufficiently complicated—a subset of a sufficiently large and nasty function space—then the result will be the existence of functions with certain properties.) One example of such an existence theorem is called the Ryll–Nardzewski fixed point theorem.

Solution 5 N.

5 N

It seems very plausible that the answer should be yes—but it turns out to be no. To get an example, consider the set that consists of the points on and above the hyperbola $y = \frac{1}{x}$ in the first quadrant of the plane, together with the origin. What is its convex hull? It is not quite the entire first quadrant. It is, in fact the open first quadrant together with the origin—and nothing else. In other words, the convex hull does not contain the positive x-axis or the positive y-axis—the origin is the only

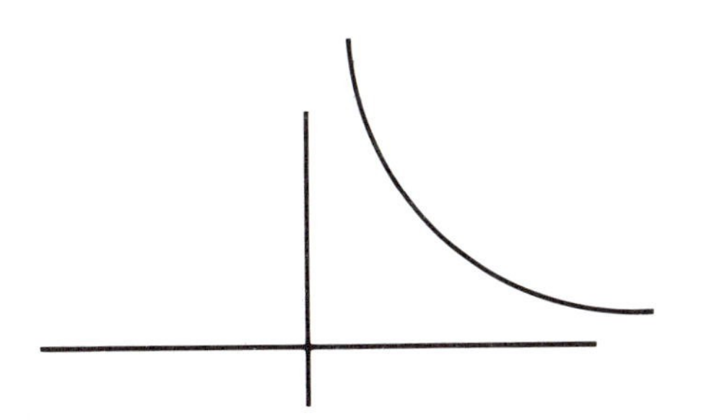

FIGURE 58

point in the axes that it does contain. Consequence: the convex hull of the given closed set is not closed.

Comment. It is legitimate to ask, and, in fact, at this point it should be asked, whether the convex hull of a compact set must be compact. The answer is yes, but that's a somewhat deeper and more specialized part of the theory. Even experts don't always know everything—and, in particular, even experts may not run across the facts in infinite-dimensional spaces, where the result just mentioned is not true.

Chapter 6. Tilings

6 A

Solution 6 A.

If the first player selects the square in the second row and second column, then after the removal of that square, and everything above and to the right of it, what is left is a large L-shaped figure, such as the unshaded part of Figure 59. From here on the magic word is symmetry:

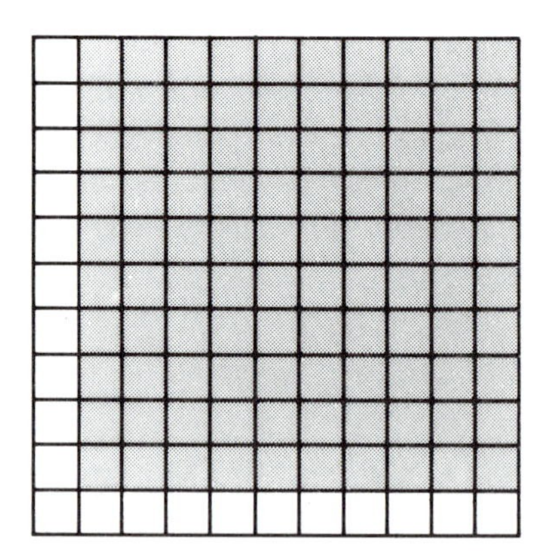

FIGURE 59

no matter what the second player does, the first player does the symmetric thing so as to symmetrize the figure, or, more explicitly, so as to produce another L-shaped configuration with equal sides. The limiting case is Figure 60 with the second player to move, and it's clear that the second player is doomed to lose.

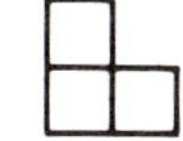

FIGURE 60

Solution 6 B.

6 B

If the first player removes the top right square, leaving a two-rowed configuration whose top row is one shorter than the bottom row, then,

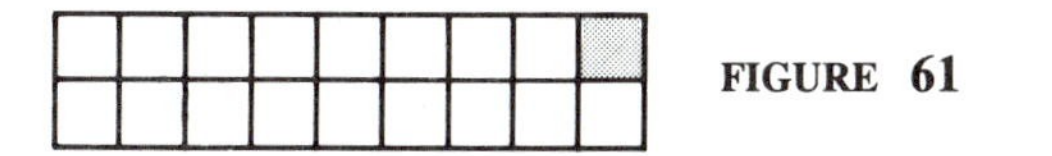

FIGURE 61

no matter how the second player answers, the first player can always reproduce such a configuration. Indeed: if the second player chooses a square in the bottom row, then, after the removal of every square above and to the right of such a square, what is left is another thin rectangle; if, however, the second player chooses one of the remaining squares in the top row, the first player can remove the square in the bottom row one step to the right of the one the second player just selected, and the "bottom = top plus one" configuration appears again. The limiting case is the configuration in Figure 62 which is just what happened in Problem 6 A, and, once again, the second player is doomed to lose.

FIGURE 62

Comment. By a pretty well established mathematical-linguistic convention, the expression "$i \times j$ rectangle" refers to a rectangle with i

rows and j columns. The present problem is about thin rectangles with 2 rows; what happens with thin rectangles that have two columns, that is with $n \times 2$ rectangles? Answer, after only a microsecond of thought: nothing happens—the same argument yields the same conclusion.

6 C

Solution 6 C.

If the first player chooses a square in the bottom row, he reduces the board to a thin chomp board of the kind considered in Problem 6 B, with the second player in the role of the first player. If the first player chooses a square in the top row, the second player can choose the square in the bottom row one stop to the right of the one the first player just selected. In either case, therefore, the second player can force the "bottom = top plus one" configuration —which is a winner.

Comment. Doesn't that seem odd —that the *second* player should be able to force a win? It's an unusual manifestation of the difference between finite and infinite.

6 D

Solution 6 D.

The most obvious fact about (finite) chomp is that it's a game in which a tie is not possible—after no more than a predictable finite number of moves, one or the other player will have to take the last square and lose.

Another fact, just a millimeter less obvious, is the existence of a first move such that, no matter how the second player answers it, the resulting situation is one that the first player also could have produced. That sounds complicated but isn't. What it means is this: if the first player removes the top right square, then whichever remaining square the second player selects was available (when the game began) as a choice for the first player also.

These two remarks yield a solution of the problem.

The assertion is that the first player can force a win. The proof is an "either, or" argument. Either the removal of the top right square enables the first player to force a win, or it does not. If it does, nothing more needs to be said. Consider the possibility that it does not. That means that if the second player faces a chomp board with the top right square removed, then the second player is *not* doomed to lose. What does "not doomed to lose" mean? It means that by playing intelligently

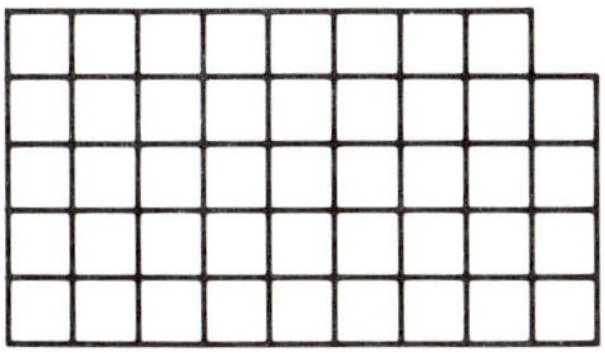

FIGURE 63

the second player can avoid becoming the loser—but, since there are no ties, that means that by playing intelligently, the second player can force a win. What does "force a win" mean? It means that the second player can make a move that has no winning answer. But, but, but—whatever such a move might be, that move was available to the first player in the first place. Either removal of the top right square is a winning move for the first player, or it is not—and if it is not then the answer to it was available as a winning move for the first player.

Comment. Isn't that a strange argument? It's correct, and it does prove that the first player can always force a win—but it doesn't give the slightest hint as to how. It is a logical argument, not an algorithmic one; it establishes the existence of a winning strategy, but not its construction. Existence proofs like that are usually called non-constructive, and they usually depend on an argument by contradiction. This one doesn't seem to—it's the most constructive non-constructive proof that I can imagine.

Solution 6 E.

6 E

Consider an equilateral triangle of side length one inch doubled by reflection through one side, each vertex marked with the name A, B, or C, of one of the three colors (see Figure 64). Assume that the answer to the question is no; then $A \neq B$, $A \neq C$, and $B \neq C$. Rotate the figure upward about the left-hand A by such an angle that the distance between the new position of the right-hand A and its old position is exactly one inch each, and once again let each vertex be marked with the name X, Y, or Z of one of the three colors (see Figure 65). The assumption implies that X and Y are different from each other and from A, so that they must be B and C (not necessarily respectively). The as-

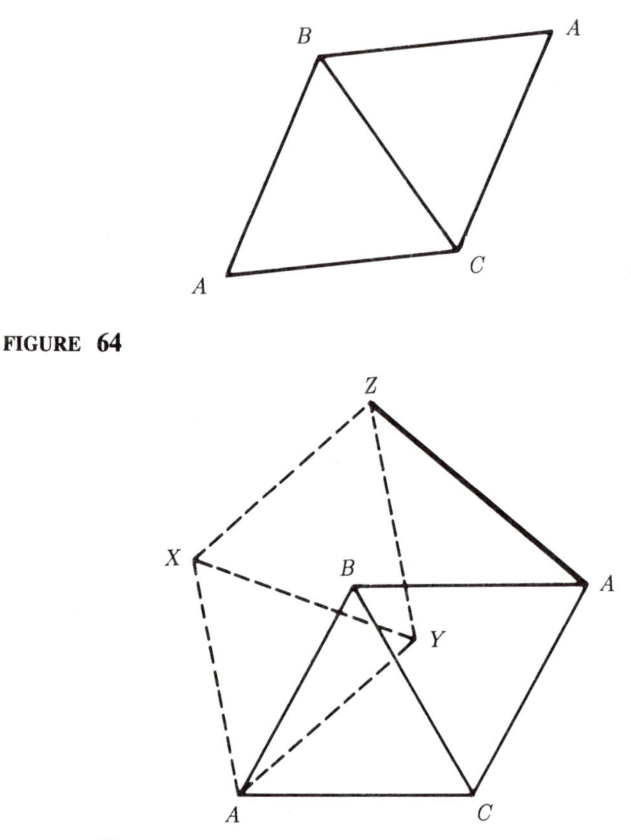

FIGURE 64

FIGURE 65

sumption therefore implies that $Z = A$; since the distance between Z and the "right" A is one inch, the contradiction has arrived.

Comment. There is another way of expressing the solution. Argue, as before, about the doubled equilateral triangle, for all possible positions of the auxiliary points B and C. Conclusion: either one of that (infinite) set of equilateral triangles contains a pair of points of the same color exactly one inch apart, or else every one of the (infinite) set of points that can play the role of the second A has the same color as the original A. The latter possibility says that every point on the perimeter of a circle (the exact value of whose radius doesn't matter very much, but is easily calculated to be $\sqrt{3}$) has the same color as the original A. Let X be any

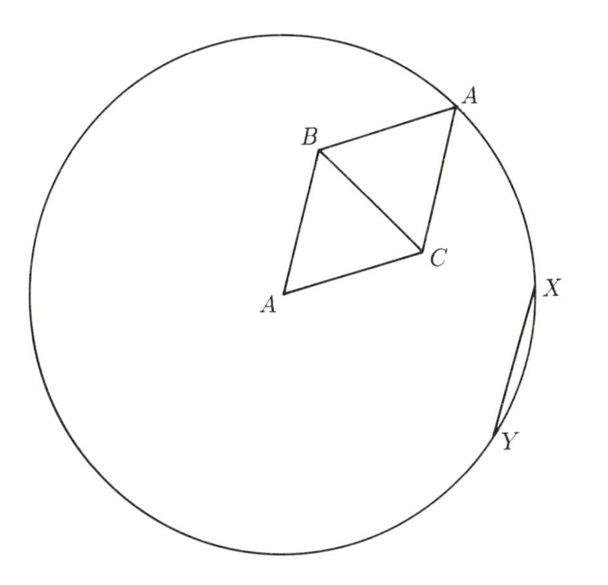

FIGURE 66

point on that circle, and let Y be the other end point of a chord of length one inch—X and Y provide an answer to the original question.

If $\mathbb{R}^2$ (the plane) in the question is replaced by $\mathbb{R}^3$, the answer becomes a trivial yes: the four vertices of a regular tetrahedron with side length one inch can have only three different colors, so that two of them must be the same.

Does the same reasoning as in the solution above yield the same conclusion for $n + 1$ colors in n-dimensional space? It is tempting to guess yes, and a careful look at the solution is likely to reinforce that guess. The guess is, in fact, correct, for all positive integer values of n, with one exception: it is not correct when $n = 1$. Indeed: color all the points of the half-closed intervals $[2n, 2n + 1)$ red ($n = 0, \pm 1, \pm 2, \ldots$), and color the complementary intervals $[2n+1, 2n+2)$ blue ($n = 0, \pm 1, \pm 2, \ldots$). Then an examination of Figure 67 indicates that no two points at a distance 1 are of the same color. What seems to be happening is that the set of distances attained by pairs of points of the same color consists of all non-negative real numbers except the odd integers.

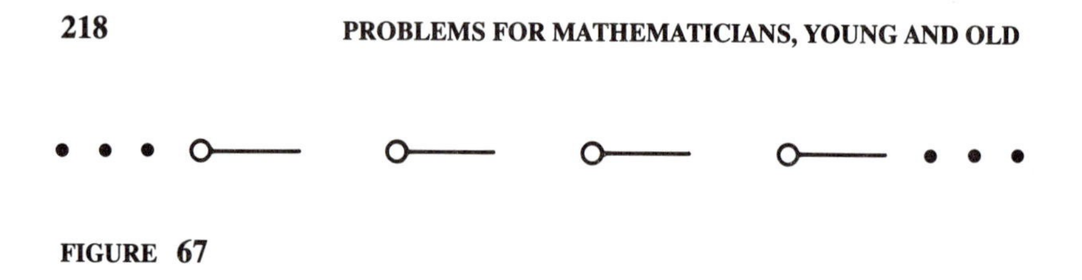

FIGURE 67

The problem belongs to what is called geometric Ramsey theory, and so do the next two; yes, the same Frank Ramsey as the one mentioned in Solution 1 B.

6 F

Solution 6 F.

There is a coloring of the plane with *seven* colors such that no two points exactly an inch apart have the same color. Just tile the plane with hexagons as in Figure 68, making the distance between two opposite vertices of each hexagon a tiny bit less than one inch—say nine-tenths of an inch. If the interior of each hexagon is colored with one of seven colors (as indicated by the numerals 1, 2, 3, 4, 5, 6, 7), then no interior point is at distance exactly one inch from another interior point of the same color, and it looks as if the assertion is proved—or almost proved.

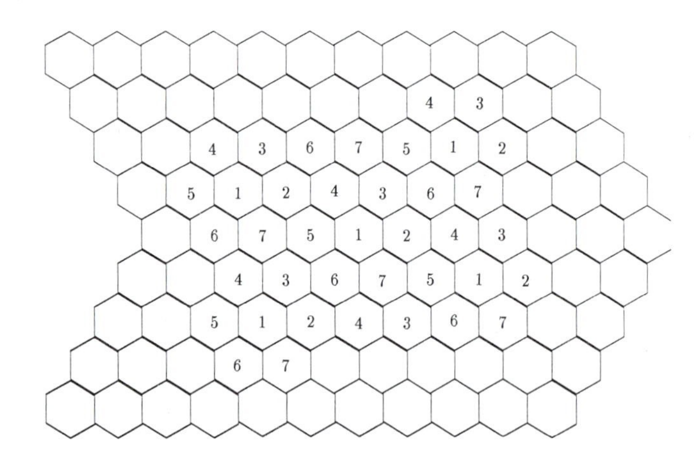

FIGURE 68

Some small fuss has to be made about the edges and vertices—to which hexagons are they to be regarded as belonging?—what are their colors? There are several intelligent ways of making the almost totally arbitrary choices. One of them is this: color the three edges of each hexagon that are at its left side, including the two vertices of the middle one of them, the same as the interior of the hexagon. That decision colors every point of the plane, and it remains true that no point is at distance exactly one inch from another point of the same color. That's it; the negative assertion is proved.

It follows, of course, that the answer for eight, or nine, or ten colors is just as negative, and it remains negative for all greater numbers of colors.

Comment. The answer for three turned out to be yes, and the answer for seven turned out to be no. It's clear, isn't it, that any time the answer is no, it remains no for all greater numbers, and any time the answer is yes, it remains yes for all smaller numbers. The results obtained so far settle the matter, therefore, for all numbers except 4, 5, and 6—what are the answers for them? Answer: nobody knows. The subject at this point has reached the frontiers of mathematics—the Ramsey questions for 4, 5, and 6 are unsolved research problems.

Solution 6 G.

6 G

If the small squares in the large one have alternating colors, as on a checkerboard, then the two deleted squares, being at two ends of a diagonal, must be of the same color. A domino, no matter how it is laid down, covers two squares of different colors. Consequence: since 50 of the 98 small squares (or 32 of the 62 in case of the standard checkerboard) that remain after the deletion are of one color and 48 (or 30) of the other, and since a domino tiling would imply that the two colors are equally represented, no such tiling is possible.

Comment. There is another way of expressing the solution that sounds different. Assign a horizontal and a vertical coordinate, running from 1 to 10, to each small square. Any two squares that a domino can cover have a total of four coordinates, whose sum is always odd. (Clear? The coordinates of the two squares are either of the form (x, y) and $(x, y+1)$ or else of the form (x, y) and $(x+1, y)$.) Consequence: the area that 49 dominoes can cover has an odd coordinate sum, whereas the removal

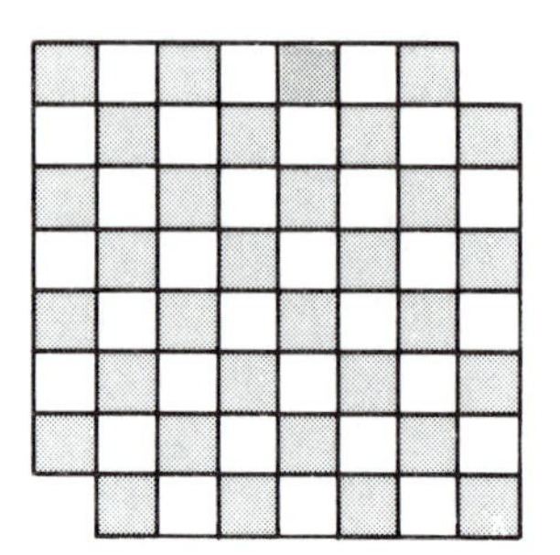

FIGURE 69

of the diagonally opposite squares $(1, 1)$ and $(10, 10)$ leaves an even coordinate sum.

6 H

Solution 6 H.

The answer is yes. Walk through the squares of the checkerboard starting with, say, $(1, 1)$, so that the walk covers every square exactly once, and ends at $(2, 1)$, right next to $(1, 1)$. There are many ways of doing that. One way is to start by walking up

$$(1, 1), (1, 2), \ldots, (1, 8),$$

then cut all the way to the right

$$(2, 8), (3, 8), \ldots, (8, 8),$$

then come down

$$(8, 7), (8, 6), \ldots, (8, 1),$$

and then go up, down, up, down, in alternating columns, ending with

$$(2, 7), (2, 6), \ldots, (2, 1).$$

The order in which the walk covers the 64 squares is a circular one; from $(2, 1)$ the walk could continue to $(1, 1)$ and repeat the whole thing. Two neighboring squares in the circle have different colors of course. Figure 70 illustrates such a walk for a 4×4 checkerboard.

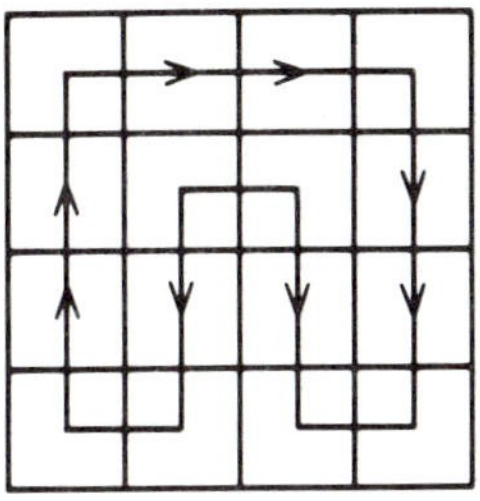

FIGURE 70

If the two deleted squares are *not* neighbors in the circle, they cut the circle into two pieces, each of which consists of an even number of squares, and each of which, therefore, can be tiled by dominoes. If the two deleted squares *are* neighbors in the circle, what they leave is just one piece consisting of an even number of squares, and therefore, it too can be tiled by dominoes. Figure 71 illustrates such a "circle" (that, to be sure, doesn't look very circular) for the 4×4 case of Figure 70. The 16 squares are the 16 squares of the checkerboard numbered in the order in which they are walked through, and, at the same time, the numbers indicate the colors: odd and even could, for instance, stand for black and white, respectively.

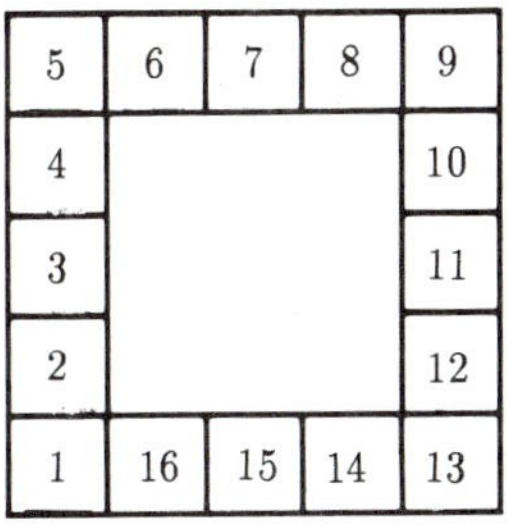

FIGURE 71

Solution 6 I.

6 I

The answer depends on which square is deleted.

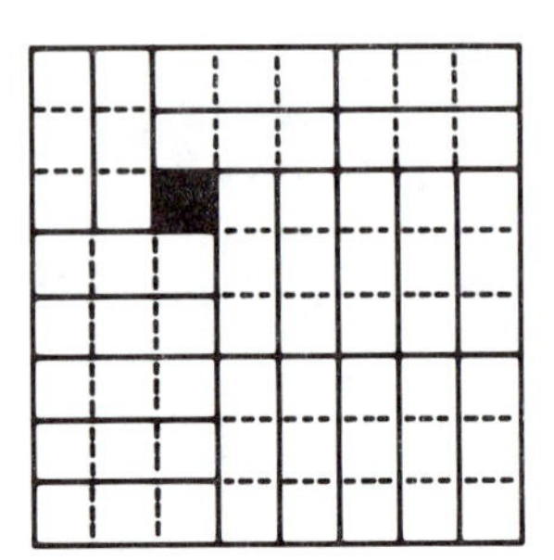

FIGURE 72

If the deleted square is such that both its coordinates are divisible by 3 (there are just four such squares), then the tiling can be done; otherwise not. The affirmative answer is proved by Figure 72.

For the negative answer, note that the sum of all the x-coordinates of the checkerboard (before the deletion) is divisible by 3 (since $1+\cdots+8$ is divisible by 3), and the same is true of the y-coordinates. If the x-coordinate of the deleted square is not divisible by 3, then neither is the sum of the remaining ones. A long domino (1×3) always covers an area the sum of whose x-coordinates is divisible by 3, and the same is true of the y-coordinates—and the negative conclusion follows.

6 J

Solution 6 J.

The diminished 8×8 checkerboard can always be tiled with corners, and, more generally, so can every $2^n \times 2^n$ checkerboard, $n = 1, 2, 3, \ldots$. Induction proofs are usually routine and dull, but the preceding sentence is a rare instance of a mathematical fact for which an approach by mathematical induction is both amusing and informative—it's easy, but fun.

For the 2×2 checkerboard the theorem is trivial—clearly every square is "omittable". To go by induction from 2^n to 2^{n+1}, locate the quadrant of the $2^{n+1} \times 2^{n+1}$ checkerboard (counted from the center point) that contains the deleted square, and use the induction hypothesis to tile that quadrant (minus the deleted square) with corners.

Delete from each of the three remaining quadrants the square next to the center of the board and use the induction hypothesis again to tile each of those quadrants (minus their deleted squares) with corners. The three deleted squares form a corner; that corner, together with all the

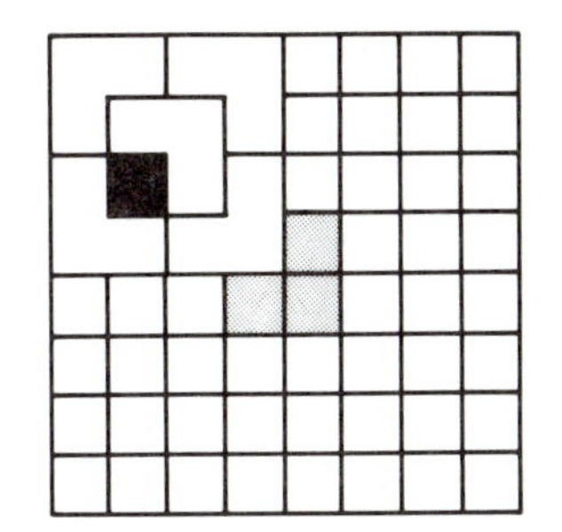

FIGURE 73

ones used in tiling the four deleted quadrants completes the corner-tiling of the whole (deleted) board.

The result of tiling a moderately large checkerboard, say for instance 16×16, with corners can be a pretty picture, such as the one in Figure 74.

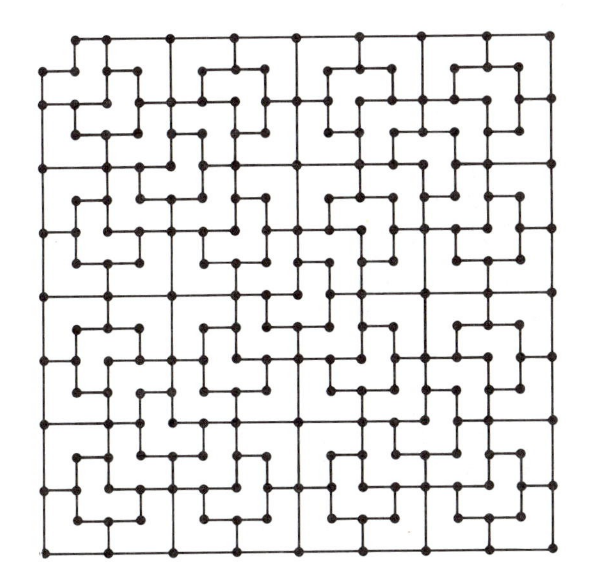

FIGURE 74

The dots at the vertices don't mean anything; they're there just to emphasize the shape of the corners.

6 K Solution 6 K.

Suppose that a large rectangle is tiled with semi-integral small ones, and form, for each tile, $(a, b) \times (c, d)$ say, the double integral

$$\int\int_{(a,b)\times(c,d)} e^{2\pi i(x+y)}\,dx\,dy.$$

That double integral is, of course, equal to the iterated integral

$$\int_a^b e^{2\pi ix}\,dx \cdot \int_c^d e^{2\pi iy}\,dy.$$

Since, by assumption, either $b - a$ or $d - c$ is an integer, it follows that one or the other of the two factors vanishes, and hence that the double integral does. Add over all tiles and infer that the corresponding double integral extended over the large rectangle also vanishes. Consequence: one of the simple integral factors of that large double integral vanishes. Recall, however, or recalculate, that

$$\int_a^b e^{2\pi ix}\,dx = \frac{1}{2\pi i}\left[e^{2\pi ib} - e^{2\pi ia}\right],$$

and hence that (if this time $(a, b) \times (c, d)$ denotes the large rectangle) either $e^{2\pi ib} = e^{2\pi ia}$ or $e^{2\pi id} = e^{2\pi ic}$. The exponential expression $e^{2\pi ix}$ is periodic with integer periods; in other words, the replacement of x by $x + k$ leaves it unchanged if and only if k is an integer. Conclusion: the large rectangle is semi-integral. The answer to the question the way it was phrased is no, or, equivalently, if a rectangle is tiled with semi-integral rectangles, then it must be semi-integral itself.

Comment. This is surely a curious solution, involving, as it does, reasoning about integers, combinatorics, geometry, and analysis—and, of course, it's the analysis that comes as a surprise. In August, 1985, Hugh Montgomery gave an invited address at the Laramie meeting of the American Mathematical Society. His official subject had nothing whatsoever to do with this rectangle problem. Just for fun, however, he began his lecture by stating the problem and presenting the solution involving the exponential integrals—having done that, he said, he could be sure that everyone understood and appreciated at least the first five minutes.

Ingenious as it may be, the exponential solution is far from the only one.

Chapter 7. Probability

Solution 7 A.

7 A

The answer is no; no such loading is possible.

Suppose, indeed, that p_i and q_i are the probabilities of face i appearing on top, $i = 1, \ldots, 6$, for the two dice—the numbers p_i and q_i must, of course, belong to the closed unit interval $[0, 1]$ to deserve to be considered as possible probabilities. Use those numbers as coefficients in forming the polynomials

$$P(x) = \sum_{i=0}^{5} p_{i+1}x^i \quad \text{and} \quad Q(x) = \sum_{i=0}^{5} q_{i+1}x^i.$$

The reason for forming these at first blush mysterious polynomials is that the way probabilities are calculated and the way polynomials are multiplied have a strong resemblance; the only difference between them is that the indices are slightly out of phase. For example: what is the probability that the sum of the two rolled dice is the third lowest it can be, which is 4, and what is the coefficient of the third lowest power of x, which is x^2, in the product of the two polynomials? Both the answers are equal to

$$p_1q_3 + p_2q_2 + p_3q_1,$$

and such equalities persist for each of the probabilities and each of the exponents. The probability question translates, therefore, to an algebra question: do there exist real numbers p_i and q_i such that all the coefficients of the product, $P(x)Q(x)$, all eleven of them, are equal to $\frac{1}{11}$?

The answer is no for a non-trivial algebraic reason, a pretty one. It cannot be true that

$$P(x)Q(x) = \frac{1}{11}\sum_{i=0}^{10} x^i,$$

and the reason is that the roots of the polynomial on the right are the non-real 11th roots of unity. (Surely everybody knows that

$$1 + x + x^2 + \cdots + x^{10} = \frac{1 - x^{11}}{1 - x}?)$$

On the other hand $P(x)$ is a polynomial of odd degree with real coefficients, and therefore $P(x)$ has at least one real root, and the same is true of $Q(x)$. Conclusion: no loading of the dice yields equiprobable sums.

Comment. There are other ways of answering the question, more elementary ways, but none so elegant. Here, for instance, is an elementary answer of a more hammer-and-tongs kind.

If there were a way of loading dice, given by $p_1, p_2, p_3, p_4, p_5, p_6$ and $q_1, q_2, q_3, q_4, q_5, q_6$, then the probabilities of the sums 2, 12, and 7 must be

$$p_1q_1, p_6q_6, \quad \text{and} \quad p_1q_6 + p_2q_5 + p_3q_4 + p_4q_3 + p_5q_2 + p_6q_1,$$

all of which must be equal to $\frac{1}{11}$. The first two of these equations say that

$$q_1 = \frac{1}{11p_1} \quad \text{and} \quad q_6 = \frac{1}{11p_6}.$$

It follows that the first and the last terms of the probability of 7 are

$$\frac{1}{11}\frac{p_1}{p_6} \quad \text{and} \quad \frac{1}{11}\frac{p_6}{p_1}.$$

Since, however, $\dfrac{x^2+y^2}{xy} \geqq 2$ is always true, it follows that

$$\frac{1}{11}\left(\frac{p_1}{p_6} + \frac{p_6}{p_1}\right) \geqq \frac{1}{11} \cdot 2 > \frac{1}{11},$$

which implies that the probability of 7 is greater than $\frac{1}{11}$—a contradiction.

7 B

Solution 7 B.

Suppose that p_i and q_i are the probabilities of face i appearing on top, $i = 1, \ldots, 6$, for the two dice, and that these probabilities are such that the probability of the sum 2 is $\frac{1}{36}$, the probability of the sum 3 is $\frac{2}{36}$, and so on all the way, through the probability $\frac{6}{36}$ for the sum 7, and up to the probability $\frac{1}{36}$ for the sum 12. Introduce, as before, the polynomials

$$P(x) = \sum_{i=0}^{5} p_{i+1}x^i \quad \text{and} \quad Q(x) = \sum_{i=0}^{5} q_{i+1}x^i,$$

and infer, as before, that

$$P(x)Q(x) = \frac{1}{36}(1+2x+3x^2+4x^3+5x^4+6x^5+5x^6+4x^7+3x^8+2x^9+x^{10}).$$

The right side is equal to

$$\frac{1}{36}(1 + x + x^2 + x^3 + x^4 + x^5)^2 = \frac{1}{36}\left(\frac{1 - x^6}{1 - x}\right)^2.$$

(That's either obvious, or, in the worst case, easy to check.) It follows that the roots of the right side are all the sixth roots of unity, except 1, each with multiplicity 2. (The reason for the multiplicity is, of course, the exponent 2.)

If ω denotes, as is customary, a primitive cube root of unity, then the sixth roots of unity other than 1 are

$$\omega,\ \omega^2,\ -1,\ -\omega,\ -\omega^2,$$

so the putative ten roots of the product $P(x)Q(x)$ are the ten numbers

$$\omega,\ \omega^2,\ -1,\ -\omega,\ -\omega^2,\ \omega,\ \omega^2,\ -1,\ -\omega,\ -\omega^2.$$

That means that five of them must be the roots of $P(x)$ and the other five the roots of $Q(x)$. How can that happen? In other words, how could those ten numbers have been sorted out into two piles of five which can serve as the roots of $P(x)$ and $Q(x)$? One way could have been

$$P(x) = (x + 1)(x - \omega)(x - \omega^2)(x + \omega)(x + \omega^2),$$

which multiplies out to $1 + x + x^2 + x^3 + x^4 + x^5$, and then $Q(x)$ is, of course, the same.

Is there any other way it could have happened? Try it. Sort out the ten candidates into piles of five, and multiply the corresponding factors so as to try to get two polynomials with real coefficients. It doesn't take much experimental time to conclude that it cannot be done in any way different from the one already mentioned—which corresponds to the classical probabilities for dice. (The "classical" probabilities are the ones for which all the faces are equally probable: $\frac{1}{6}$, $\frac{1}{6}$, $\frac{1}{6}$, $\frac{1}{6}$, $\frac{1}{6}$, $\frac{1}{6}$.) Conclusion: the only way of loading a pair of dice in such a way that the probability of the occurrence of each sum from 2 to 12 is the same as it is for honest dice is the classical way, or, otherwise expressed, there is no dishonest way of loading honest dice.

7 C Solution 7 C.

The answer is yes—which is not obvious at the start. Here is one way: let the numbers of dots on the faces of one die be

$$1, 2, 2, 3, 3, 4,$$

and

$$1, 3, 4, 5, 6, 8$$

on the other. The check that every sum from 2 to 12 inclusive has the same probability as always is routine, but fun in a way.

Comment. Is this solution an accident? How could it have been discovered? Are there any other solutions?

No, it's not an accident, and yes, there is a natural but rather boring way to discover it—and following that way through shows that there are no other solutions. Here is an outline of how the dedicated seeker could have proceeded.

Each die must have a face with only one dot—otherwise the sum 2 could never be achieved—and it can't have more than one—otherwise the sum 2 would have higher probability than $\frac{1}{36}$. Since the sum 3 must be achieved, one of the dice must have two dots on at least one face. Believe it or not, once that much has been said, that is, once the dot distribution has begun with

$$1, 2, ?, ?, ?, ?$$

and

$$1, ?, ?, ?, ?, ?,$$

the rest is forced. To illustrate the kind of forcing that goes on, ask this question: could both dice have a face with two dots on it? In other words, is something like

$$1, 2, ?, ?, ?, ?$$

and

$$1, 2, ?, ?, ?, ?$$

possible? Sure, it's possible—the usual dot distribution is an instance of it. If a different distribution is wanted, then that second 2 leads to trouble, as follows. As things stand, the probability of a sum 3 is already $\frac{2}{36}$; that implies that no other 2 can ever be used. Since the sum 4 must be achieved, one die must have three dots on some face. Keep looking, keep exhausting cases, keep counting possibilities and probabilities. The process is not a thrill, but the conclusion it leads to is the "trouble" mentioned: the conclusion is that once that second 2 was put down on the second die, the dot distribution has to be the ordinary one.

Solution 7 D.

7 D

Yes, it's possible, and it isn't even complicated. Here is one way: put two 5's on A and four 2's, let die B have two 1's and four 4's, and let die C have three dots on every face. (Equivalently: load the dice, which are spotted as dice always are, so that on die A the 5-spot has probability $\frac{1}{3}$ of showing and the 2-spot has probability $\frac{2}{3}$, on die B the one-spot has probability $\frac{1}{3}$ and the four has probability $\frac{2}{3}$, and, finally, on die C the three-spot has probability 1.)

The calculations that verify non-transitivity are easy. What, for instance, is the probability that $A > B$? How can it happen that $A > B$? Answer: that will happen if and only if A shows 5 (probability $\frac{1}{3}$) or else A shows 2 and B shows 1 (probability $\frac{2}{3} \cdot \frac{1}{3}$), so that the total answer is $\frac{1}{3} + \frac{2}{9} = \frac{5}{9}$. What about the probability that $B > C$? That's easy: $B > C$ can happen if and only if B shows 4, which has probability $\frac{2}{3}$. The calculation of $C > A$ is equally trivial.

Solution 7 E.

7 E

The answer is no; it is quite possible for the slower horse to be the more frequent winner. Here is a possibility—perhaps it is not a realistic one on racetracks, but it illustrates the mathematics of the problem without getting bogged down in irrelevant arithmetic.

You watch five races. In two of them X runs the course in 1.2 minutes, and in the three others he takes 1.3 minutes, 1.4 minutes, and 1.5 minutes. The horse Y, on the other hand, runs only one race each in 1.2 minutes, and in 1.3, and in 1.4, and he runs two races in 1.5 minutes. Let $P(X \leq t)$ indicate the percentage of races that X runs in time t or less, and, similarly, let $P(Y \leqq t)$ indicate the same information about Y. A tabular summary of your observations looks like this:

Time t	$P(X \leqq t)$	$P(Y \leqq t)$
1.2	.40	.20
1.3	.60	.40
1.4	.80	.60
1.5	1.00	1.00.

The table says that X is the faster horse: for each t (except the extreme 1.5) he ran more races in time t or less than did Y.

It is, nevertheless, dangerous to jump to conclusions. These observations could have come about, for instance, by X being a fast starter, which temporarily frightens Y, although Y is in fact the the better race horse and will win most of the races between them.

But just exactly what happened in those five races that you watched? It could have happened that the times that it took X and Y to run them compared as in this table:

X	Y
1.2	1.3
1.2	1.4
1.3	1.5
1.4	1.5
1.5	1.2.

In that case it looks like it would be wise to bet on X in the future: he won 80 percent of the races. But the same observations about percentages could have come from a record such as this:

X	Y
1.2	1.5
1.2	1.5
1.3	1.2
1.4	1.3
1.5	1.4,

and in that case X may not be the smart choice. In this latter case X won 60 percent of the races, and, just by juggling the numbers a bit, and allowing more races, it is possible to arrange matters so that although X is "faster" (in the sense that his distribution function is always greater than that of Y), X loses 99 percent of the races he runs against Y.

Solution 7 F.

7 F

The issue is settled if the toss of the n coins results in one head and $n-1$ tails or, the other way around, in one tail and $n-1$ heads. The probability that some particular one of the n gamblers gets heads and all the others tails is $\frac{1}{2^n}$, and, consequently, the probability that a toss results in one head and $n-1$ tails is $\frac{n}{2^n}$. The probability that a toss results in one tail and $n-1$ heads is, of course, the same, $\frac{n}{2^n}$, and, therefore, the probability of the issue being settled by any particular toss is

$$\frac{n}{2^n} + \frac{n}{2^n} = \frac{2n}{2^n}.$$

For $n = 3$ this probability is $\frac{3}{4}$, and that suggests that it's not likely that many tosses are needed to choose the winner; for $n = 4$ the answer is $\frac{1}{2}$, and that is still quite reasonable. For large n, say for instance for $n = 10$, the probability is $\frac{20}{1024}$, which is less than .02—the chances are less than one in fifty of settling the issue in any one toss, and the procedure is likely to result in discouragement.

Solution 7 G.

7 G

The probability that I win might be guessed on the basis of a symmetry principle: I am as likely to win as any of my three opponents, no more and no less so, and "therefore" the answer must be $\frac{1}{4}$. Is that reasoning correct?

The reasoning by symmetry may be correct, but, at the very least, it needs justification.

1. One possible justification uses an "infinite argument", as follows.

If the first toss results in HTTT or in THHH (with my coin being the one that is listed first), then I win. The probability of one of those things happening is $\frac{1}{16} + \frac{1}{16}$ $(= \frac{1}{8})$.

If they don't happen, I still have a chance to win. I could, for instance, win on the second toss, provided that nobody won on the first, which happens exactly when the number of heads that came up on the first toss is even. The probability of four heads is $\frac{1}{16}$, the probability of zero heads is $\frac{1}{16}$, and the probability of two heads (which can happen in $\binom{4}{2}$ ways each with probability $\frac{1}{16}$) is $\frac{6}{16}$; it follows that the probability of an even number of heads is $\frac{1}{16} + \frac{1}{16} + \frac{6}{16} = \frac{1}{2}$. Consequence: the probability of my winning on the second toss, which is the probability that nobody wins on the first multiplied by the probability of my winning the second, is $\frac{1}{2} \cdot \frac{1}{8}$.

If I don't win on either of the first two tosses, I still have a chance of winning, by, for instance, winning on the third, provided that nobody won on the first two. The probability of all that happening is $\frac{1}{2} \cdot \frac{1}{2} \cdot \frac{1}{8}$.

An inductive machine is running here; what it says is that the probability of my winning on (exactly) the kth toss is

$$\left(\frac{1}{2}\right)^{k-1} \cdot \frac{1}{8}.$$

Conclusion: there are infinitely many different ways that I could win, and the total probability of my winning is the sum of their probabilities, which is

$$\sum_{k=1}^{\infty} \left(\frac{1}{2}\right)^{k-1} \cdot \frac{1}{8} = \frac{1}{8} \cdot 2 = \frac{1}{4}.$$

2. Another possible justification uses a simple recursion equation, as follows. Let x be the (unknown) probability of my winning. Begin by arguing as above that the probability of my winning on the first toss is $\frac{1}{8}$ and the probability of nobody winning on the first toss is $\frac{1}{2}$. Then continue by saying that there are two ways I can win: either I win on the first toss, with probability $\frac{1}{8}$, or nobody wins on the first toss and I go on to win, a combination of events that has probability $\frac{1}{2} \cdot x$. (That's right, isn't it? If a second toss becomes necessary to settle the issue, then, since the coins have no memory, we are back in the same situation as before the first toss—the probability of my winning is still the same x.) Consequence:

$$x = \frac{1}{8} + \frac{1}{2} \cdot x.$$

Conclusion:

$$x - \frac{1}{2} \cdot x = \frac{1}{2} \cdot x = \frac{1}{8} \quad \text{and} \quad x = \frac{1}{4}.$$

Solution 7 H.

7 H

For three players the answer is easy, and the idea should be plain from Solution 7 G. What's proved there, by a "justified symmetry argument", is that if four of us toss for a winner by "odd man out", then the probability that I win is $\frac{1}{4}$. The reasoning is perfectly general—it applies to n people, for any positive integer n, just as well as to 4. Conclusion: if three of us toss for winner by "odd man out", then the probability that I win is $\frac{1}{3}$. Can this game be changed to one between two people?

The answer is pretty obviously yes, via an inelegant and artificial construction—but it is a construction just the same. Idea: double one of the players. Differently said: I toss the coin, then you toss it, and then you toss it again. If the three tosses result in one heads and two tails, or one tails and two heads, then the person who tossed the odd one wins; otherwise the coin has to be tossed (three times) again. (A different description of the same game might be that I toss one coin and you toss two, but since the problem asked for a game using *one coin*—"an" honest coin—that might be considered unfair.)

There is still another way of achieving the same result, a way that doesn't look so inelegant or unfair. Namely: you and I toss a coin alternately; the first one to toss heads wins. What then is the probability that I win? How can I win? I can win only if you begin by tossing tails (with probability $\frac{1}{2}$). If that does happen, then either I toss heads (and I have won); or I toss tails, you toss tails, and I toss heads (and I have won); or I toss tails, you toss tails, I toss tails, you toss tails, and I toss heads (and I have won); and so on ad infinitum. The probability that I win on the second toss is $\frac{1}{4}$ (your tails and my heads); the probabililty that I win on the fourth toss is $\frac{1}{4} \cdot \frac{1}{4}$ (your tails, my tails, your tails, my heads); and so on. The total probability that I win is therefore

$$\sum_{n=1}^{\infty} \left(\frac{1}{4}\right)^n = \frac{1}{3}.$$

A different proof that this game works uses a simple recursion equation (compare Solution 7 G) as follows. Let x be the (unknown) probability of my winning. Then the probability that I lose, which is $1-x$, is the probability that I lose on the first toss (you get heads), with probability $\frac{1}{2}$, plus the probability that I do not lose on the first toss (you get tails), and then *you* lose. At that point, however, I am in the situation that you were in before the game began—and it follows that the probability of your losing is the same as the probability of my winning, which is x. In other words

$$1 - x = \frac{1}{2} + \frac{1}{2}x.$$

That equation is easy to solve; the solution is $x = \frac{1}{3}$.

This is not the first time that an "infinite argument" has been used to get the answer, and some people worry about that. Surely the problem asked for a *finite* game, whatever that may mean, and surely the game here described is not a finite one. Well, that depends. What do we mean by a finite game? The strongest answer is that the game ends in a predictable finite number of steps—such as heads you win, tails you lose, and the game is over in one toss. In that interpretation of finite, almost none of the popular gambling games is finite. Certainly "odd man out" is not—that could go on ad infinitum—and neither is craps, the popular dice game. What is true is that all these games are "locally finite". It is not necessarily true that such a game must end in one step, or a hundred steps; the probability that the game ends in any prescribed finite number of steps, such as a hundred, is *not* equal to 1. What is true, however, is that with probability equal to 1 the entire game will take only a finite number of steps—and that's good enough in the real world for both gamblers and casinos.

7 I Solution 7 I.

It should be clear that the number $\frac{1}{\pi}$ has nothing to do with the issue; the question is really one about an arbitary number (in the unit interval). The answer is yes.

Given a number α between 0 and 1, let

$$\alpha = .\alpha_1\alpha_2\alpha_3\ldots$$

be its binary (base 2) expansion, so that

$$\alpha = \sum_{n=1}^{\infty} \frac{\alpha_n}{2^n},$$

where each α_n is 0 or 1, and then play the game as follows. Keep tossing the coin, recording the result, say β_n, of the nth toss as 0 when it falls tails and 1 when it falls heads; you win the game if the first time that $\alpha_n \neq \beta_n$ your result, β_n, is 0 (and therefore $\alpha_n = 1$). In other words, you win if the first time the β sequence differs from the α sequence, the β sequence is smaller.

The motivation for this procedure is to think of α as determining the interval $(0, \alpha)$, and think of the coin tosses as determining a number in $[0, 1]$; "win" means "land in the prescribed interval".

How can you win? To see the answer, it helps to introduce some notation: let $n_1, n_2, n_3, \ldots$, be the positions of the 1's that occur in the sequence α. One way to win is to toss so that the β sequence agrees with the α sequence for the first $n_1 - 1$ terms and then disagrees. The probability of that is $\frac{1}{2^{n_1}}$ which is equal to the partial sum up to n_1 of the binary representation of α. The next way to win is to toss so that the β sequence agrees with the α sequence for the first $n_2 - 1$ terms and then disagrees. The probability of that is $\frac{1}{2^{n_2}}$ which is equal to the number obtained from the partial sum up to n_2 of the binary representation of α by replacing the first 1 with a 0. These two ways of winning are the beginning of an infinite sequence. The (infinite) sum of the corresponding probabilities is exactly the binary representation of α.

Comment. Since

$$\frac{1}{3} = .010101\ldots$$

(binary expansion), this solution is in harmony with Solution 7 H (where $\alpha = \frac{1}{3}$). In the game there described, you win if and only if the first head comes up on an even numbered toss.

Solution 7 J.

7 J

If, to look at an extreme, the distribution is

(50 white, 0 black) and (0 white, 50 black),

then the probability of drawing a white ball is

$$\frac{1}{2} \cdot 1 + \frac{1}{2} \cdot 0 = \frac{1}{2}.$$

If, for a different example, the distribution is

(25 white, 25 black) and (25 white, 25 black),

then the probability is

$$\frac{1}{2} \cdot \frac{1}{2} + \frac{1}{2} \cdot \frac{1}{2} = \frac{1}{2}.$$

If these two answers suggest anything at all, they might suggest that the answer is always $\frac{1}{2}$, independently of how the balls are distributed—but that's not so.

It does make a difference how they are distributed. If, for instance, the distribution is

(50 white, 50 black) and (0 white, 0 black),

then the probability is

$$\frac{1}{2} \cdot \frac{1}{2} + \frac{1}{2} \cdot 0 = \frac{1}{4}.$$

One of the right questions to ask in circumstances such as these is this: how should the balls be distributed so as to make the probability of drawing a white ball as large as possible? The answer is intuitively plausible:

(1 white, 0 black) and (49 white, 50 black),

but the computation that proves this answer to be indeed optimal is rather cumbersome.

7 K

Solution 7 K.

Here is a "theological" answer. If you keep breaking yardsticks, the average of the lengths of the pieces you keep is likely to be half a yard. In other words, the expected value of a number chosen at random from the unit interval is $\frac{1}{2}$. That makes it seem likely that the number of times you have to choose numbers so as to have them add up to 1 or more is

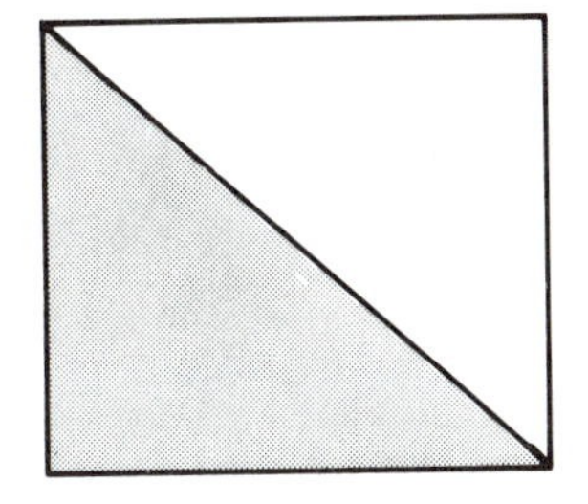

FIGURE 75

more than two and less than three. There is only one interesting number between 2 and 3; the answer *must* be e.

For a mathematical answer, integral calculus comes to the rescue. To begin somewhere near the beginning, what is the probability that the sum of two randomly chosen numbers x_1 and x_2 is *less* than 1? Geometrically the questions is this: what is the measure (area) of the part of the unit square in the (x_1, x_2) plane where $x_1 + x_2 < 1$? The answer is obviously $\frac{1}{2}$, but that's not the most useful observation; the useful thing is to write the answer as

$$\int_0^1 dx_1 \int_0^{1-x_1} dx_2.$$

In probability language this iterated integral is the answer to the question: what is the probability that the sum of two random numbers between 0 and 1 is less than 1? The answer suggests (correctly), that the corresponding answer for three numbers (whether the question is asked in geometry or in probability) is

$$\int_0^1 dx_1 \int_0^{1-x_1} dx_2 \int_0^{1-x_1-x_2} dx_3$$

an integral whose value is $\frac{1}{6}$. The induction is obvious: the probability that the sum of n random numbers in the unit interval is less than 1 is

$$\int_0^1 dx_1 \int_0^{1-x_1} dx_2 \int_0^{1-x_1-x_2} dx_3 \cdots \int_0^{1-x_1-\cdots-x_{n-1}} dx_n \cdots,$$

an integral whose value is $\frac{1}{n!}$.

It follows that the probability that n is the smallest number of x's needed to add up to *at least* 1 is

$$\text{Prob}\ (x_1 + \cdots + x_n \geqq 1) - \text{Prob}\ (x_1 + \cdots + x_{n-1} \geqq 1)$$
$$= \left(1 - \frac{1}{n!}\right) - \left(1 - \frac{1}{(n-1)!}\right) = \frac{n-1}{n!}.$$

Once that's known, the calculation of the expectation is a routine exercise in infinite series: the expectation is the sum over all possible values of the product of each value with its probability. In other words, the answer is

$$\sum_{n=2}^{\infty} n \cdot \frac{n-1}{n!} = \sum_{n=0}^{\infty} \frac{1}{n!} = e,$$

which is the answer that the theological argument gave.

7 L

Solution 7 L.

Consider the following possible strategy. You draw a number, and a second, and a third, and so on—no matter what you see, you keep drawing numbers up to and including the 50th. You look at each number, but no matter what it is you don't stop. Then, on your 51st draw, you compare the result with the first fifty: if that draw is greater than all the ones before, you stop right there. If it is not, then you know you would lose if you stopped—the preceding fifty beat the number you now have—so you keep drawing. If the 52nd number beats the first fifty, you stop with it; if not you keep going. To put it simply: keep drawing till you beat the first fifty. Of course that may never happen—it is quite possible that the largest number in the hat was among the first fifty. And even if that is not so, you are far from certain of winning with this strategy—it is quite possible that the 51st number beats all of its predecessors but just isn't the largest—the 73rd number, had you but kept drawing that long, is in fact the largest. No matter—here is a definite strategy. What is the probability that you will win with it?

To calculate the answer would not be a particularly pleasant chore, and, fortunately, it is not necessary; the answer to the original question can be learned without that calculation. What is obviously true is that IF it happens that the second largest number is among the first fifty that you drew, and IF it happens that the largest is among the second fifty, then, for sure, the strategy will result in your winning. You could win even if that is not the way the largest and second largest happen to lie—but that way is for sure. All right—what is the probability that the second largest number is among the first fifty and the largest among the second fifty?

It is clear that the probability of each of those events considered separately is $\frac{1}{2}$ ($=\frac{50}{100}$), and it is tempting to say that therefore the probability that they both occur is $\frac{1}{4}$ ($=\frac{1}{2}\cdot\frac{1}{2}$). The temptation should be resisted. The answer it gives is not right—it is a good approximation, but it is not right. The trouble is that the two events (second largest among first fifty, largest among second fifty) are not independent—the usual reason for being permitted to multiply probabilities does not apply. But the correct answer is easy enough to calculate: all we have to decide is the conditional probability that, given that the second largest number is among the first fifty draws, the largest number is among the second fifty. If the second largest is already known to be among the first fifty, then there are 99 other equally likely draws that might conceivably yield the largest, and 50 among them are the ones among the second fifty—in other words, the conditional probability sought is $\frac{50}{99}$. The probability of the for sure winning distribution is $\frac{1}{2}\cdot\frac{50}{99}$, which is obviously greater than $\frac{1}{2}\cdot\frac{50}{100}$ (and is, in fact, equal to .2525...).

Conclusion: with a good strategy the probability of your winning is at least $\frac{1}{4}$—which implies that a payoff of $4.00 would be favorable, more than fair, and that, consequently, with a payoff of $5.00 you would be robbing the bank.

Comment. The problem of computing the exact probability has been considered in the mathematical literature for a long time. It is sometimes embedded into a story about a businessman needing to hire a secretary and interviewing a hundred applicants one after another—it has come, therefore, to be recognized under the name "the secretary problem".

The strategy outlined in the solution is of the right kind, but it is not the best. The best strategy is to keep drawing, as indicated, but not to go on to $\frac{100}{2}$ draws—the correct number is $\frac{100}{e}$. That is: if the hat contains not a hundred numbers, but n, then draw $\frac{n}{e}$ of them (or as close as you can come to that fraction), without stopping, and then keep drawing till you have beaten all those draws. As n becomes large, the strategy gets closer and closer to being optimal, and the probability of your winning gets closer and closer to the maximal probability, which is $\frac{1}{e}$. The number 2 is a very rough approximation to e, and the answer .2525... is a very rough approximation to $\frac{1}{e}$.

Some light is shed on the situation by an examination of a small value of n, namely $n = 4$. There are four distinct numbers in the hat; what should you do? Answer: decide on a number d of slips to draw and discard before you start trying to beat the discarded ones. The probability p of winning is given by the table

d:	0	1	2	3
p:	$\frac{6}{24}$	$\frac{11}{24}$	$\frac{10}{24}$	$\frac{6}{24}$.

The best strategy is to choose $d = 1$.

Chapter 8. Analysis

8 A Solution 8 A.

There is really only one sensible interpretation of the three dots that indicate that the tower should be continued indefinitely, namely as a limit. What is meant is that we should form finite towers such as

$$\sqrt{2}^{\sqrt{2}^{\sqrt{2}^{\cdot^{\cdot^{\cdot^{\sqrt{2}^{\sqrt{2}}}}}}}},$$

make them higher and higher, and then try to let the heights tend to infinity. More explicitly: write $x_1 = \sqrt{2}$, and then, for every positive integer n, define x_{n+1} as $\sqrt{2}^{x_n}$; the question is whether or not the sequence $\{x_n\}$ converges, and, if so, to what. The natural guess would seem to be no—how could it converge?

One question is obviously relevant: what kind of a function of x does the expression $\sqrt{2}^x$ define? Answer: a monotone increasing function. That is: if $x < y$, then $\sqrt{2}^x < \sqrt{2}^y$. Isn't that clear? If there is any doubt about it, take the logarithm of both sides of the inequality, or, more precisely, look at the result of forming that logarithm, agree that it's a correct statement, and then form its exponential.

Two consequences follow from the monotone increasing character of $\sqrt{2}^x$. One is that the sequence $\{x_n\}$ is increasing (an obvious induction), and the other is that the sequence is bounded from above by 2. To see the latter consequence, replace the topmost $\sqrt{2}$ in the tower

$$\sqrt{2}^{\sqrt{2}^{\sqrt{2}^{\cdot^{\cdot^{\cdot^{\sqrt{2}^{\sqrt{2}}}}}}}}$$

by 2, thus getting a larger number, and observe that the result telescopes downward. (That is, replace what are now the top two exponents, namely $\sqrt{2}^2$, by the value they give, namely 2, and then continue downward.) These two consequences imply the conclusion: the sequence $\{x_n\}$ is convergent to some limit less than or equal to 2.

Can that limit be evaluated? Sure—easy. Call it t, so that

$$t = \lim_n x_n.$$

It follows that

$$\sqrt{2}^t = \sqrt{2}^{\lim_n x_n} = \lim_n \sqrt{2}^{x_n} = \lim_n x_{n+1} = \lim_n x_n = t,$$

or, throwing away the steps and keeping only the conclusion,

$$\sqrt{2}^t = t.$$

This equation can be solved by inspection: it has the obvious solutions 2 and 4. The latter is not possible, because we already know that $t \leqq 2$—the conclusion is that $t = 2$, and that's that.

Comment. This problem, and related ones, have been studied extensively for many years; I was once shown an unpublished bibliography of well over a hundred items. The main question is what happens when $\sqrt{2}$ is replaced by an arbitrary positive number: for which numbers does the surprising convergence continue, and, when it does, what is the limit? In other words, given a positive number x, write $x_1 = x$, and, for every positive integer n define x_{n+1} as x^{x_n}; for which x's can the resulting sequence $\{x_n\}$ converge? The answer turns out to be that the sequence converges if and only if

$$\frac{1}{e^e} \leqq x \leqq e^{1/e}.$$

8 B

Solution 8 B.

The absolute continuity and the end point conditions imply that

$$f(x) = \int_0^x f'(t)\,dt \quad \text{and} \quad f(x) = -\int_x^1 f'(t)\,dt$$

for every x in $[0,1]$. Add these two equations and make use of the obvious inequality between the integral of a function and its absolute value; the result is

$$2|f(x)| \leqq \int_0^x |f'(t)|\,dt + \int_x^1 |f'(t)|\,dt = \int_o^1 |f'(t)|\,dt.$$

Consequence (in view of the assumed value of the rightmost term):

$$|f(x)| \leqq \frac{1}{2}$$

for every value of x. Conclusion: it is not possible that $f(\frac{1}{2}) = 1$ or -1, or 2.

Comment. The method of reaching the conclusion does not exclude the values 0 and $\frac{1}{2}$; they are "possible" in the sense of not having been proved to be impossible. The argument leaves open, however, the question of whether they are really possible, that is, whether there really exist functions f satisfying the conditions and such that $f(\frac{1}{2}) = 0$ or $f(\frac{1}{2}) = \frac{1}{2}$. An efficient way of settling the issue is to exhibit functions that do these "possible" things, and that is easy enough. If $f(x) = \frac{1}{4}\sin 2\pi x$, then $f(\frac{1}{2}) = 0$, and if $f(x) = \frac{1}{2}\sin \pi x$, then $f(\frac{1}{2}) = \frac{1}{2}$, and each of these functions has all the desired properties.

8 C

Solution 8 C.

The answer is no: it is possible to construct a uniformly continuous and continuously differentiable function f whose derivative is not bounded. A precise epsilontic construction would be a bit of a bore; a set of geometric instructions might serve just as well.

The whole infinite line is irrelevant; let us pay attention to the positive half only. That is: define $f(x)$ to be 0 when $x < 0$, and then start going to the right. Begin with the value 0, make f have slope 1 for a while, and then remain constant for a while; then make it have slope 2 for a short time, and then constant for a long time; then make it have slope 3 for a very short time, and then constant for a very long time; etc. The "short times" are to be arranged so as to make f bounded, and the "long times" so as to guarantee that the domain of f is the entire line. It is necessary, of course, to pay attention to smoothness during the construction; as the slope of the graph is changed from a large value to 0 the corners must be rounded.

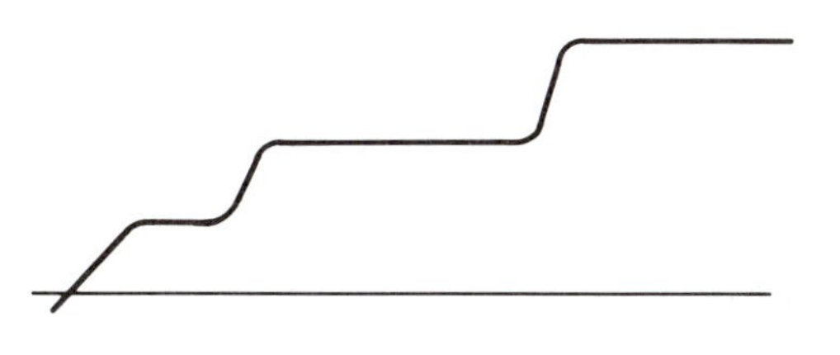

FIGURE 76

Comment. If the "long times" are left out, then f can be made continuous on a finite interval, but in that case the derivative f' will be infinite at the right end.

Solution 8 D.

8 D

There are many ways to learn that the answer is yes; here is one. Start with the fact that $x < e^x$ for every positive x. (In fact the inequality holds for all real x, but for the moment that is irrelevant.) Then deduce, one after another, the following relations:

$$|\log x| < x \quad (1 < x < \infty),$$

$$|\log \sqrt[4]{x}| < \sqrt[4]{x} \quad (1 < x < \infty),$$

$$|\log \frac{1}{\sqrt[4]{x}}| < \frac{1}{\sqrt[4]{x}} \quad (0 < x < 1),$$

$$\frac{1}{4}|\log x| < \frac{1}{\sqrt[4]{x}} = x^{-1/4} \quad (0 < x < 1),$$

$$|\log x|^2 < 16x^{-1/2} \quad (0 < x < 1).$$

Since $-\frac{1}{2} > -1$, all is well; the function $|\log|^2$ is dominated by a function that is integrable, and, consequently, it itself is integrable.

8 E Solution 8 E.

There is a pretty argument by symmetry, as follows. Under the change of variables that replaces x by $\frac{1}{y}$ the integral

$$\int_0^\infty \frac{\log x}{1+x^2}\,dx$$

becomes

$$\int_\infty^0 \frac{-\log y}{1+\frac{1}{y^2}} \cdot \frac{(-dy)}{y^2},$$

which simplifies to

$$-\int_0^\infty \frac{\log y}{1+y^2}\,dy.$$

Consequence: the unknown integral is equal to its own negative, and, therefore, has the value 0.

The argument may be pretty, but, of course, it is illegitimate—at least till it's proved to make sense. It still remains to be decided whether or not the integrand is integrable. The part from 0 to 1 causes no trouble; the function log itself is integrable there, and its quotient by something greater than 1 is all the more so. As for the part from 1 to ∞, one way to dispose of it is to invoke de l'Hôpital's rule and infer that $\frac{\log\sqrt{x}}{\sqrt{x}}$ tends to 0 as x tends to ∞. Consequence: $\log x \leqq 2\sqrt{x}$ for x sufficiently large, and therefore

$$\frac{\log x}{1+x^2} \leqq \frac{2\sqrt{x}}{1+x^2} < \frac{2\sqrt{x}}{x^2} = 2x^{-3/2}$$

for x sufficiently large. Since the last term is integrable, so is the first, and everything is settled.

8 F Solution 8 F.

It happens that there is a simple characterization of all possible universal chords—a surprising characterization. Namely: a number c is a universal chord if and only if it is the reciprocal of a positive integer. The answer to the questions posed is, therefore, that $\frac{1}{2}$ is a universal chord, and so are .1 $(= \frac{1}{10})$, and .2 $(= \frac{1}{5})$, but .3 and $\frac{1}{e}$ are not.

The easy part is the proof that if $c = \frac{1}{n}$ with $n = 1, 2, 3, \ldots$, then c is a universal chord. Suppose, indeed, that f is one of the functions under consideration, and consider the telescoping sum

$$\begin{aligned}\left(f\left(\frac{1}{n}\right)-f\left(\frac{0}{n}\right)\right) &+\left(f\left(\frac{2}{n}\right)-f\left(\frac{1}{n}\right)\right)+\left(f\left(\frac{3}{n}\right)-f\left(\frac{2}{n}\right)\right)\\ &+\cdots+\left(f\left(\frac{n}{n}\right)-f\left(\frac{n-1}{n}\right)\right)\\ &= f(1)-f(0)=0.\end{aligned}$$

If one of the n summands vanishes, then we're done: $\frac{1}{n}$ is a chord of f. If that doesn't happen, then in the interval $\left[0, 1-\frac{1}{n}\right]$ the expression

$$d(x) = f\left(x+\frac{1}{n}\right) - f(x)$$

attains at least one positive value and at least one negative value. Clear? The reason is that the sum of the n values that the difference function d takes on at the n points

$$\frac{0}{n}, \frac{1}{n}, \frac{2}{n}, \ldots, \frac{n-1}{n}$$

is 0, and, by present assumption, none of those n values is 0—so that there must be a positive one and a negative one among them. But d is continuous; if it's positive somewhere and negative somewhere, then it must be 0 somewhere—and that means that $\frac{1}{n}$ is a chord of f.

Suppose now that a number c is *not* the reciprocal of a positive integer; it is to be proved that c cannot be a universal chord. In other words, it is to be proved that there exists at least one function f of the kind under study such that c is not a chord of f.

As the first step toward the construction of such a function f, let us construct a continuous function g on $[0, 1]$ such that

$$g(0) = 0, \qquad g(1) = 1,$$

and g is periodic with period c. The last part means that

$$g(x+c) = g(x)$$

whenever $0 \leqq x \leqq 1 - c$. It's easy enough to construct such a function just by drawing its graph, but, if desired, an explicit formula can be given for one—for instance

$$g(x) = \frac{\sin\left(\frac{2\pi x}{c}\right)}{\sin\left(\frac{2\pi}{c}\right)}.$$

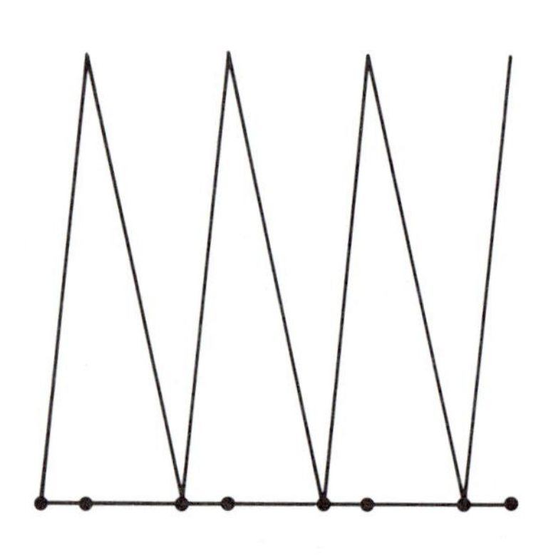

FIGURE 77

The second step finishes the job: just put

$$f(x) = x - g(x).$$

To check that f is an admissible function is easy enough: f is continuous, $f(0) = 0$, and $f(1) = 0$. Could c be a chord of f? That is: could there be a value of x such that $0 \leqq x \leqq 1 - c$ and such that $f(x + c) = f(x)$? No, there couldn't. Reason:

$$f(x + c) - f(x) = ((x + c) - g(x)) - (x - g(x)) = c.$$

8 G

Solution 8 G.

Suppose that a function f is Cesàro continuous at a single point x_0. Replacement of $f(x)$ by $g(x) = f(x) - f(x_0)$ justifies the assumption that the value of the function at the point x_0 in question is 0. Replacement of $g(x)$ by $h(x) = g(x + x_0)$ justifies the assumption that the crucial point x_0 in question is 0. These two replacements together help quite a bit to simplify the notation, and they lose no generality; from now on it will be assumed that the function f is Cesàro continuous at 0 and $f(0) = 0$.

If a, b are arbitrary real numbers, then the sequence

$$\{a, b, -(a+b), a, b, -(a+b), a, b, -(a+b), \ldots\}$$

is Cesàro convergent to 0, and, therefore, by the assumed Cesàro continuity, so is the sequence

$$\{f(a), f(b), f(-(a+b)), f(a), f(b), f(-(a+b)), f(a), f(b), f(-(a+b)), \ldots\}.$$

Consequence:

$$f(-(a+b)) = -(f(a) + f(b)) \tag{1}$$

for all a and b. (The verification of that consequence is a tiny exercise—it is a good test of one's understanding of the definition of Cesàro convergence.) With $b = 0$ the equation (1) yields the information that

$$f(a) = -f(-a). \tag{2}$$

A change of notation will now put matters into a more natural context: replace a and b by $-a$ and $-b$ in (1) and use (2); the result is

$$f(a+b) = f(a) + f(b) \tag{3}$$

for all a and b.

What we have reached is a celebrated functional equation (sometimes called the Hamel equation), and the question of which functions f can satisfy that equation has been studied extensively. Here is a quick outline of the first steps of those studies. Replace b by a to infer that $f(2a) = 2f(a)$, and replace b by $-a$ to infer that $f(-a) = -f(a)$. Iterate the result to infer that $f(pa) = pf(a)$ for every integer p. Replace a by $\frac{a}{q}$, where q is an arbitrary non-zero integer, and infer that $f(\frac{a}{q}) = \frac{1}{q}f(a)$. Consequence: $f(ra) = rf(a)$ for every rational number r. Conclusion: the only *continuous* functions that can satisfy (3) are the ones given by $f(x) = Ax$.

That doesn't quite solve the problem at hand: the function f was not assumed to be continuous, but Cesàro continuous. Suppose, however, that $\{x_n\}$ is an arbitrary convergent sequence of real numbers (convergent in the usual sense), with limit 0. Is there a sequence whose averages are exactly the x's? That is: is there a sequence $\{y_n\}$ such that

$$\frac{y_1 + y_2 + \cdots + y_n}{n} = x_n$$

for every n? Sure, just solve the indicated equations. The equations demand that

$$y_1 + y_2 + \cdots + y_n = nx_n \qquad \text{for } n = 1, 2, \ldots$$

and

$$y_1 + y_2 + \cdots + y_{n-1} = (n-1)x_{n-1} \qquad \text{for } n = 2, 3, \ldots .$$

Consequence:

$$y_n = nx_n - (n-1)x_{n-1} \qquad \text{for } n = 1, 2, 3, \ldots$$

(with some sensible agreement about how to interpret x_{1-1}—the simplest probably being 0). Since

$$\frac{y_1 + y_2 + \cdots + y_n}{n} \to 0,$$

it follows from the assumed Cesàro continuity of f that

$$\frac{f(y_1) + f(y_2) + \cdots + f(y_n)}{n} \to 0.$$

In view of the additivity property of f this says that

$$f(x_n) = f\left(\frac{y_1 + y_2 + \cdots + y_n}{n}\right) \to 0.$$

What just happened? Answer: from the assumption that $x_n \to 0$ it was deduced that $f(x_n) \to 0$—in other words f was proved to be continuous at 0. The additivity property of f can be brought in again: it implies that the conclusion can be translated from 0 to any other point. In other words: from the assumption that f is Cesàro continuous at 0, it follows that f is continuous everywhere.

That settles everything: f is not only additive, but also continuous, and therefore $f(x) = Ax$ for some constant A. The conclusion depends on the normalizations made at the beginning. When they are unnormalized, the result is that if f is a function that is Cesàro continuous at 0, then f must have the form $f(x) = Ax + B$ for some constants A and B.

The question was whether x^2 and $\sqrt{x}$ were Cesàro continuous, and the answer is no—they are not of the form $Ax + B$. Cesàro continuity turned out to be a stronger condition than could reasonably have been expected—much stronger, for instance, than continuity.

Solution 8 H.

8 H

The natural way to find the length of a curve is to approximate it by polygons that are as simple as possible and to calculate their lengths. As a first approximation consider the curve that agrees with the Cantor function in the middle and joins the two ends of the middle segment to the two corners of the unit square as indicated in Figure 78. The base of each triangle is $\frac{1}{2} - \frac{1}{2}\mu_1 = \frac{1}{2}(1 - \mu_1)$; the altitude is $\frac{1}{2}$. The hypotenuse of each triangle is

$$\sqrt{\left(\frac{1}{2}\right)^2 + \left(\frac{1}{2}(1-\mu_1)\right)^2} = \frac{1}{2}\sqrt{1 + (1-\mu_1)^2}.$$

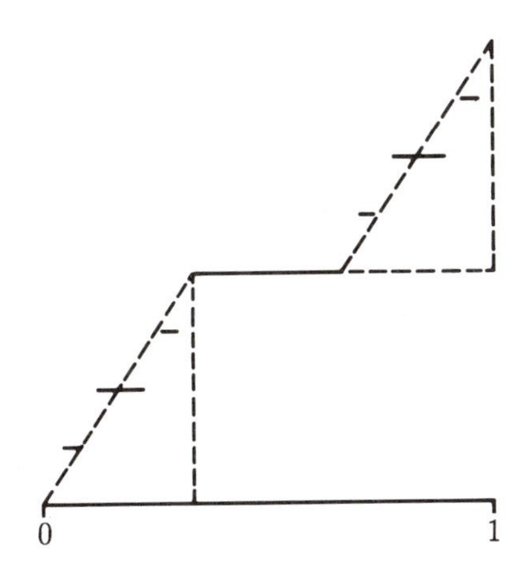

FIGURE 78

There are two such hypotenuses, and, therefore, the length of the first approximation is

$$\mu_1 + \sqrt{1 + (1-\mu_1)^2}.$$

The second approximation is in Figure 79. For each triangle the base is

$$\frac{1}{4}(1-\mu_1) - \frac{1}{2}\mu_2 = \frac{1}{4}(1 - \mu_1 - 2\mu_2),$$

the altitude is $\frac{1}{4}$, and the hypotenuse is

$$\sqrt{\left(\frac{1}{4}\right)^2 + \left(\frac{1}{4}(1-\mu_1-2\mu_2)\right)^2} = \frac{1}{4}\sqrt{1+(1-\mu_1-2\mu_2)^2}.$$

The length of the second approximation is

$$\mu_1 + 2\mu_2 + \sqrt{1+(1-\mu_1-2\mu_2)^2}.$$

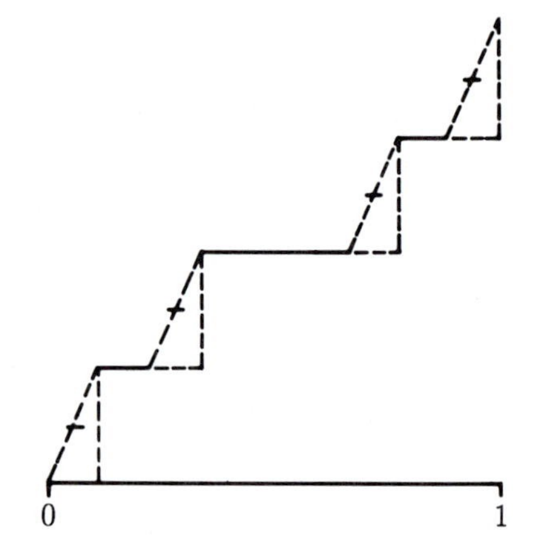

FIGURE 79

The inductive procedure is clear; the length of the nth approximation must be

$$\mu_1 + 2\mu_2 + \cdots + 2^{n-1}\mu_n + \sqrt{1+(1-\mu_1-2\mu_2-\cdots-2^{n-1}\mu_n)^2}.$$

The limit is

$$1-\alpha+\sqrt{1+(1-(1-\alpha))^2} = 1-\alpha+\sqrt{1+\alpha^2}.$$

Comment. Isn't it curious and non-obvious that the answer is independent of the μ's?—it depends on α only. When $\alpha = 0$ the answer is 2.

Solution 8 I.

8 I

The fact is that if f is an entire function that is not a translation, then the equation $f(f(z)) = z$ always has a solution. A possible proof is to assume that f is an entire function for which the equation $f(f(z)) = z$ has no solution and hence no fixed point and try to prove that it must be a translation.

The assumption about no fixed points implies that the equation

$$g(z) = \frac{f(f(z)) - f(z)}{z - f(z)}$$

makes sense (because the denominator is never zero) and is different from 0 (because the numerator is never zero) for every value of z.

Can the function g attain the value 1? If so, if, for instance, $g(z) = 1$, then $f(f(z)) - f(z) = z - f(z)$, or $f(f(z)) = z$—a contradiction. Consequence: there are at least two complex numbers (namely 0 and 1) that the entire function g does not take on as values. At this point it is possible to use the high powered Picard theorem which says, exactly under these conditions, that that can happen only if the function g is a (non-zero) constant. That is: there exists a complex number $c \neq 0$ such that

$$f(f(z)) - f(z) = cz - cf(z)$$

for all z.

Consequence:

$$z = \frac{1}{c}(f(f(z)) - f(z) + cf(z))$$

for all z. This equation produces a further unexpected consequence: it implies that the function f is univalent. (That is: it cannot happen that $f(u) = f(v)$ for two different complex numbers u and v.) Univalent entire functions are not easy to come by: the only ones are of the form

$$f(z) = az + b \quad (a \neq 0).$$

This is obvious for polynomials; for other entire functions it constitutes another use of Picard's theorem.

One more use of the assumption that there are no fixed points finishes the argument: if $a \neq 1$, then $az + b$ always has a fixed point (the equation $az + b = z$ is solvable). Conclusion: $f(z)$ is a translation.

8 J

Solution 8 J.

If p is a monic polynomial,

$$p(z) = z^n + \sum_{j=0}^{n-1} a_j z^j$$

(or, in the extreme case, $p(z) = 1$), that maps the unit disk into itself, then p is a monomial ($p(z) = z^n$, $n = 0, 1, 2, \ldots$).

There are at least two proofs, and they both have merit.

One of the proofs is phrased in the spirit of complex analysis, as follows. The assumption implies that

$$\frac{p(z)}{z^n} = 1 \qquad \text{at } \infty,$$

and hence, by the maximum modulus theorem, $\frac{p(z)}{z^n}$ must be a constant.

Another proof is more "real" in spirit. Form $|p|^2$ and integrate it with respect to normalized Lebesgue measure μ on the perimeter of the unit circle (that is the "arc length" measure that assigns the value 1 to the entire perimeter). The result is that

$$\int |p(z)|^2 \, d\mu(z) = 1 + \sum_{j=0}^{n-1} |a_j|^2.$$

If $a_j \neq 0$ for any j, then this mean value (that is, the integral) is strictly greater than 1, which implies that $|p|$ takes values strictly greater than 1.

8 K

Solution 8 K.

Is it a surprise to learn that every number in the unit interval is a semi-universal chord?

It is indeed true that for every number c in $[0, 1]$ and for every continuous function f that vanishes at 0 and at 1, either c or $1 - c$ must be a chord of f. An elegant proof begins by extending the function f to the entire line by requiring that it be periodic of period 1. Given c, consider the (extended) graphs of $f(x)$ and $f(x+c)$, and infer, from the assumption that $f(0) = f(1) = 0$ that those graphs must have a point of intersection, and, in fact, in view of periodicity, that they must have a point of intersection whose x coordinate, say x_0, is in $[0, 1]$. That means, of course, that $f(x_0 + c) = f(x_0)$—but, careful, that doesn't yet mean that c is a chord of f, because it may not be true that $x_0 + c \leqq 1$. If, however, $x_0 + c > 1$, then $y_0 = x_0 + c - 1 \leqq 1$, and y_0 does something pleasant. Namely:

$$\begin{aligned} f(y_0 + 1 - c) &= f(x_0) \quad \text{(by the definition of } y_0\text{)} \\ &= f(x_0 + c) \quad \text{(by the defining property of } x_0\text{)} \\ &= f(x_0 + c - 1) \quad \text{(by the periodicity of } f\text{)} \\ &= f(y_0), \end{aligned}$$

so that, victory!, $1 - c$ is a chord of f.

Comment. The technique of this proof yields an alternative approach to proving that the reciprocals of the positive integers are universal chords, as follows. The result just proved implies that $\frac{1}{2}$ is a chord of every f (because $\frac{1}{2} = 1 - \frac{1}{2}$). Proceed by induction. Either $\frac{1}{3}$ or $\frac{2}{3}$ $(= 1 - \frac{1}{3})$ is a chord of f. If it's $\frac{1}{3}$, we're done; if it's $\frac{2}{3}$, we can indulge in a small bit of trickery. The trick is to realize that in the proof above what was important was that f was periodic of some period p (the fact that the p was 1 was irrelevant), and the conclusion was that one or the other of the numbers obtained by dividing p into two parts in the ratio of c and $1 - c$ (that is, one of the numbers cp and $(1 - c)p$) had to be a chord of f. Once that's granted, then the information that $\frac{2}{3}$ is a chord of f, together with what was already proved about $\frac{1}{2}$, implies that $\frac{1}{3} = \frac{1}{2} \cdot \frac{2}{3}$ is a chord of f. Keep going: either $\frac{1}{4}$ or $\frac{3}{4}$ $(= 1 - \frac{1}{4})$ is a chord of f; if it's $\frac{1}{4}$, we're done, but if it's $\frac{3}{4}$, then apply the information about $\frac{1}{3}$, the same way as the information about $\frac{1}{2}$ was just applied, to infer that $\frac{1}{4}$ is a chord too—and so on ad infinitum.

Solution 8 L.

8 L

This is a trick question. Many people are likely to answer it with an impatiently contemptuous yes; a possible detailed argument goes like this. If $p(x)$ is a polynomial, then $|p(x)| \to \infty$ as $x \to \pm\infty$. The assumption that $p(x)$ is bounded from below implies that the behavior $p(x) \to -\infty$ is ruled out, and hence it must be the case that $p(x) \to +\infty$ as $x \to \pm\infty$. That implies that $p(x)$ must be very large and positive outside a sufficiently large closed interval. Inside such an interval, p is a real-valued continuous function, and, as such, it necessarily attains its greatest lower bound.

The trickiness comes from the interpretation of the word polynomial. The proof just given assumes that it refers to a polynomial in one variable, but, as we all know, the same word can be and is used to refer to polynomials in several variables. Once that is realized, the negative answer to the question becomes a possibility, and here, indeed, is a concrete example:

$$p(x, y) = x^2 + (xy - 1)^2.$$

Clearly $p(x, y) \geqq 0$. Since the y-axis ($x = 0$) and the hyperbola ($xy = 1$) have no points in common, $p(x, y)$ never attains the value 0. If, however, $\varepsilon > 0$, and if $x = \varepsilon$ and $y = \frac{1}{\varepsilon}$, then $p(x, y) = \varepsilon^2$; this proves that 0 is actually the (greatest) lower bound.

8 M

Solution 8 M.

The answer is yes, but differentiability is a red herring: in fact if a function on the real line can be uniformly approximated by polynomials, then it must be a polynomial. Suppose, indeed, that f is a function and that $\{p_n\}$ is a sequence of polynomials that converges uniformly to f. It follows that $|p_n(x) - p_m(x)| < 1$ for all x whenever n and m are sufficiently large, and hence, in particular,that $p_n(x) - p_m(x)$ is uniformly bounded for n and m sufficiently large. Consequence (since $p_n(x) - p_m(x)$ is a polynomial): if n and m are sufficiently large, then $p_n(x) - p_m(x)$ is a constant, and that implies that $f = \lim p_n$ is a polynomial.

8 N

Solution 8 N.

Let S be the set of all squares and F the set of all square-free positive integers. Since every positive integer is the product of a square and a square-free number, it follows that

$$\left(\sum_{\substack{n=1 \\ n\in S}}^{k} \frac{1}{n}\right) \cdot \left(\sum_{\substack{n=1 \\ n\in F}}^{k} \frac{1}{n}\right) \geqq \left(\sum_{n=1}^{k} \frac{1}{n}\right).$$

The right-hand term tends to infinity with k, whereas the first factor on the left side remains bounded as k becomes large; it follows that the second factor on the left side tends to infinity. A computational trick can now be used to show that

$$\left(\sum_{\substack{n=1\\ n\in P}}^{k} \frac{1}{n}\right)$$

tends to infinity, as follows:

$$\exp\left(\sum_{\substack{n=1\\ n\in P}}^{k} \frac{1}{n}\right) = \prod_{\substack{n=1\\ n\in P}}^{k} \exp\left(\frac{1}{n}\right) > \prod_{\substack{n=1\\ n\in P}}^{k} \left(1+\frac{1}{n}\right) \geqq \left(\sum_{\substack{n=1\\ n\in F}}^{k} \frac{1}{n}\right).$$

Since the last term of this chain becomes infinite, so does the first.

Comment. It is a corollary of the result just proved that there are infinitely many primes.

If a subseries of the harmonic series happens to converge, then the complementary subseries must diverge, but, caution!, there is no reason to believe the converse. The reciprocals of the primes are a case in point: they constitute a subseries of the harmonic series that diverges, but that fact by itself gives no information about the complementary subseries. Do the reciprocals of the composite numbers constitute a convergent or divergent series? The answer is obvious: since every even number greater than 2 is composite, and since the reciprocals of the even numbers cannot constitute a convergent series (because half their sum would be the sum of the harmonic series), it follows that the reciprocals of the composite numbers constitute a divergent series.

Solution 8 O.

8 O

Every number of the form $e(\{\varepsilon_n\})$ is irrational; the proof in general is just like in the special case in which all the ε's are equal to $+1$. Suppose, indeed, that $e(\{\varepsilon_n\}) = \frac{p}{q}$ for some integers p and q with $q > 0$. Multiply through by $q!$ to get

$$0 = \left(\frac{p}{q} - \sum_{n=0}^{\infty} \frac{\varepsilon_n}{n!}\right) q! = k - \sum_{n=q+1}^{\infty} \frac{\varepsilon_n \cdot q!}{n!},$$

where k is an integer. This turns out to be a contradiction; a calculation shows that

$$\sum_{n=q+1}^{\infty} \frac{\varepsilon_n \cdot q!}{n!}$$

is never 0 and is always strictly less than 1 in modulus.

Here is the calculation:

$$\begin{aligned}\sum_{n=q+1}^{\infty} \frac{q!}{n!} &= \frac{1}{q+1} + \frac{1}{(q+1)(q+2)} + \frac{1}{(q+1)(q+2)(q+3)} + \cdots \\ &< \frac{1}{q+1} + \frac{1}{(q+1)^2} + \frac{1}{(q+1)^3} + \cdots \\ &= \frac{1}{q} \leqq 1,\end{aligned}$$

and therefore

$$\sum_{n=q+2}^{\infty} \frac{q!}{n!} = \sum_{n=q+1}^{\infty} \frac{q!}{n!} - \frac{1}{q+1} < \frac{1}{q} - \frac{1}{(q+1)} = \frac{1}{q(q+1)}.$$

Consequence:

$$\left| \sum_{n=q+1}^{\infty} \frac{\varepsilon_n \cdot q!}{n!} \right| \leqq \sum_{n=q+1}^{\infty} \frac{q!}{n!} < 1$$

and

$$\left| \sum_{n=q+1}^{\infty} \frac{\varepsilon_n \cdot q!}{n!} \right| \geqq \left| \frac{1}{q+1} - \sum_{n=q+2}^{\infty} \frac{q!}{n!} \right| > \frac{1}{q+1} - \frac{1}{q(q+1)} \geqq 0.$$

Corollary: e $(= e(\{1, 1, 1, \ldots\}))$ is irrational.

8 P Solution 8 P.

The answer is yes. To prove that

$$\sum_{n=1}^{\infty} \frac{\varepsilon_n}{n} = e$$

is possible, argue (using the divergence of the harmonic series) as follows. Choose $\varepsilon_1 = +1$, $\varepsilon_2 = +1$, etc., just long enough to have the partial sum exceed e. Then choose the next block of ε_n's to be -1 just long enough till the partial sum goes below e. Continue alternating this way; convergence is implied by the relation $\frac{1}{n} \to 0$.

Comment. Does the number e really have anything to do with either the statement or the proof?

Solution 8 Q.

8 Q

The series $\sum_{n\in A} \frac{1}{n}$ converges.

For the proof, consider the positive integers

$$1, \ldots, 9$$

$$11, \ldots, 19, 21, \ldots, 29, \ldots, 91, \ldots, 99$$

$$111, \ldots, 119, 121, \ldots, 129, \ldots, 991, \ldots, 999$$

$$\ldots$$

in whose decimal representation 0 does not occur. The number of terms in the nth row is 9^n, and, therefore, by a crude estimate, the sum of the reciprocals of the nth row is less than or equal to $9^n \cdot \frac{1}{10^{n-1}}$. It follows that the total sum is less than or equal to

$$\sum_{n=1}^{\infty} \frac{9^n}{10^{n-1}} = 9 \cdot \sum_{n=0}^{\infty} \left(\frac{9}{10}\right)^n = 9 \cdot \frac{1}{1 - \frac{9}{10}} = 90.$$

Comment. The question can be asked for every radix r in place of 10; the corresponding estimate is

$$\sum_{n=1}^{\infty} \frac{(r-1)^n}{r^{n-1}} = r(r-1).$$

The simplest radix is, of course, 2; in that case the bound is 2. The integers whose reciprocals are to be added up are (not in decimal notation but in the obviously appropriate dyadic notation)

$$1, 11, 111, 1111, \ldots.$$

In decimal notation these are

$$1, 3, 5, 7, \ldots, 2^n - 1, \ldots.$$

What in fact is the sum $\sum_{n=1}^{\infty} \frac{1}{2^n-1}$ of their reciprocals? A simple "closed" formula for it doesn't seem to exist, but easy estimates on the tail of the series make arbitrarily good approximations available; the answer is (in decimal notation) $1.60669515\ldots$.

Back to decimals: what is the value of the sum $\sum_{n \in A} \frac{1}{n}$? The necessary estimates are considerably more subtle in that case; the answer they yield is $23.1034\ldots$.

Chapter 9. Matrices

9 A

Solution 9 A.

The answer is no: a real vector space cannot be the union of a finite number of proper subspaces.

Suppose indeed that $\mathbb{V}$ is a real vector space, and that

$$\mathbb{V} = \mathbb{S}_1 \cup \cdots \cup \mathbb{S}_n,$$

where, of course, the $\mathbb{S}$'s are proper subspaces of $\mathbb{V}$. Since, in particular, $\mathbb{S}_1$ is a proper subspace, n cannot be 1.

Assume now, with no loss of generality, that no $\mathbb{S}_j$ is included in the union of the others—if it were it could just be omitted. Let x_0 be a vector that is not in $\mathbb{S}_1$, and let x_1 be a vector that is in $\mathbb{S}_1$ but is not in the union $\mathbb{S}_2 \cup \cdots \cup \mathbb{S}_n$ of the other $\mathbb{S}$'s. Assertion: the set, call it $\mathbb{L}$, of all vectors of the form $x_0 + \alpha x_1$ (where α is an arbitrary real number) is disjoint from $\mathbb{S}_1$. Reason: if $x_0 + \alpha x_1$ were equal to a vector y in $\mathbb{S}_1$, then it would follow that $x_0 = y - \alpha x_1$ is in $\mathbb{S}_1$, contrary to the assumption about x_0.

The set $\mathbb{L}$ has nothing in common with $\mathbb{S}_1$; how much can it have in common with the other $\mathbb{S}_j$'s? Answer: not more than one vector each. The reason is this: if it were the case that α and β are two distinct real numbers such that both $x_0 + \alpha x_1$ and $x_0 + \beta x_1$ are in some $\mathbb{S}_j$, then it would follow that

$$(\alpha - \beta)x_1 = (x_0 + \alpha x_1) - (x_0 + \beta x_1) \in \mathbb{S}_j,$$

contrary to the choice of x_1.

All the work is over; the time has come to draw conclusions. The set $\mathbb{R}$ of all real numbers is infinite, and, therefore, the set $\mathbb{L}$ of vectors is infinite. The set $\mathbb{L}$, however, has only a finite number of vectors in common with the union $\mathbb{S}_1 \cup \cdots \cup \mathbb{S}_n$, which implies that that union cannot cover all of $\mathbb{V}$.

Comment. The main property of $\mathbb{R}$ that was needed in the proof was that $\mathbb{R}$ is infinite—the proof works just as well for every infinite field. For vector spaces over a finite field, however, the answer to the original question is yes. If, for instance, $\mathbb{V}$ is the 2-dimensional vector space over $\mathbb{Z}/2$ that consists of four vectors 0, x, y, and $x+y$, then $\{0, x\}$ is a proper subspace, and so also are $\{0, y\}$ and $\{0, x+y\}$—and their union is $\mathbb{V}$. People who know about (real) Banach spaces might be interested to note that such spaces can't even be unions of a countable infinity of proper subspaces—the reason is the Baire category theorem.

If you know the difference between countable sets and uncountably infinite ones, re-examine the proof and discover that it proves something stronger. Since $\mathbb{R}$ is not only infinite, but, in fact, uncountable, the proof proves that a real vector space cannot be the union of countably many proper subspaces. That argument does not work for countable fields (such as the field of rational numbers)—but it works just as well for every uncountable field.

Solution 9 B.

9 B

The answer is yes: if $\mathbb{S}_1, \ldots, \mathbb{S}_n$ are subspaces of a finite-dimensional $\mathbb{V}$, all having the same dimension, then they necessarily have a simultaneous complement.

If the common dimension of the $\mathbb{S}$'s is equal to the dimension of $\mathbb{V}$, which is an awkward way of saying that all the $\mathbb{S}$'s are equal to $\mathbb{V}$, then the trivial subspace (the zero subspace) $\mathbb{O}$ is a simultaneous complement, and the situation is totally uninteresting. We might as well assume that the $\mathbb{S}$'s are proper subspaces. In that case it is an immediate consequence of Solution 9 A that the union of the $\mathbb{S}$'s is not equal to the whole space $\mathbb{V}$. Consequence: there exists a vector x_0 that does not belong to any of the $\mathbb{S}$'s, and it follows that the set of all scalar multiples of x_0 (the span of x_0) is disjoint from all the $\mathbb{S}$'s. In other words: there exists a 1-dimensional space $\mathbb{T}$ disjoint from all the $\mathbb{S}$'s.

Is there a 2-dimensional space with that property—or a 3-dimensional one? What, in fact, is the largest possible dimension that a subspace can have and still be disjoint from all the $\mathbb{S}$'s? Since the dimension of $\mathbb{V}$ is finite, the question must have an answer; call it m, and let $\mathbb{T}$ be a subspace of dimension m that is disjoint from all the $\mathbb{S}$'s. Assertion: $\mathbb{T}$ is a simultaneous complement of all the $\mathbb{S}$'s.

How could it fail? Could it, for instance, happen that $\mathbb{T}$ is not a complement of $\mathbb{S}_1$—which means that $\mathbb{S}_1 + \mathbb{T}$ is not equal to $\mathbb{V}$? If that happened, then $\mathbb{S}_1 + \mathbb{T}, \mathbb{S}_2, \ldots, \mathbb{S}_n$ would be proper subspaces, and consequently there would exist a vector z that does not belong to any of them. It would follow that the span of z and $\mathbb{T}$ is a subspace strictly larger than $\mathbb{T}$ that is disjoint from each of $\mathbb{S}_1, \ldots, \mathbb{S}_n$, which contradicts the maximality of m.

Comment. The comment about countability in Solution 9 A can be made here too: the proof proves that countably many subspaces of a finite-dimensional $\mathbb{V}$, all having the same dimension, necessarily have a simultaneous complement.

9 C

Solution 9 C.

Yes, they both have square roots. All that is needed to find them is a little sensible computation. It is not necessary to solve nine horrible equations in nine unknowns—the game can be won by looking at matrices that (are assumed to) have a lot of zeroes and solving two easy equations in two unknowns. Here, for the record are the results:

$$\begin{pmatrix} 0 & x & 1 \\ 0 & 0 & \frac{1}{x} \\ 0 & 0 & 0 \end{pmatrix}^2 = \begin{pmatrix} 0 & 0 & 1 \\ 0 & 0 & 0 \\ 0 & 0 & 0 \end{pmatrix}$$

and

$$\begin{pmatrix} 0 & 1 & x \\ 0 & 0 & 0 \\ 0 & \frac{1}{x} & 0 \end{pmatrix}^2 = \begin{pmatrix} 0 & 1 & 0 \\ 0 & 0 & 0 \\ 0 & 0 & 0 \end{pmatrix}.$$

Comment. Warning: the language is not meant to imply that these results (even with a free floating parameter x) are unique.

Solution 9 D.

9 D

The answer is yes.

Observe, to begin with, that if the characteristic polynomial of A is

$$\alpha_0 + \alpha_1\lambda + \cdots + \alpha_{n-1}\lambda^{n-1} + \lambda^n,$$

then $\alpha_0 \neq 0$. (Reason: since A is invertible, 0 is not an eigenvalue of A.) By the Hamilton-Cayley equation

$$\alpha_0 + \alpha_1 A + \cdots + \alpha_{n-1}A^{n-1} + A^n = 0,$$

and therefore

$$\alpha_0 A^{-1} + \alpha_1 + \cdots + \alpha_{n-1}A^{n-2} + A^{n-1} = 0.$$

Consequence:

$$A^{-1} = p(A),$$

where

$$p(\lambda) = \frac{1}{\alpha_0}(-\alpha_1 - \cdots - \alpha_{n-1}\lambda^{n-2} - \lambda^{n-1}).$$

Solution 9 E.

9 E

The only way a linear transformation A on an n-dimensional vector space can have $n+1$ eigenvectors with the property that every subset of n of them is linearly independent is to be a scalar multiple of the identity.

Suppose, indeed, that $f_1, \ldots, f_n, f_{n+1}$ are eigenvectors of A such that any n of them are linearly independent, with corresponding eigenvectors $\lambda_1, \ldots, \lambda_n, \lambda_{n+1}$. Since any n of the f's are linearly independent, it follows that any n of them span the whole space. Each f, therefore, is a linear combination of the others; in particular

$$f_{n+1} = \sum_{i=1}^{n} \alpha_i f_i.$$

Note that none of the α's can be 0. Reason: if $\alpha_i = 0$, then the complement of f_i within $\{f_1, \ldots, f_n, f_{n+1}\}$ is linearly dependent.

Apply A to the equation to get

$$Af_{n+1} = \sum_{i=1}^{n} \alpha_i \lambda_i f_i.$$

Since, however,

$$Af_{n+1} = \lambda_{n+1} f_{n+1} = \sum_{i=1}^{n} \alpha_i \lambda_{n+1} f_i,$$

it follows (from the linear independence of $\{f_1, \ldots, f_n\}$) that

$$\alpha_i \lambda_i = \alpha_i \lambda_{n+1} \qquad \text{for each } i = 1, \ldots, n.$$

Since $\alpha_i \neq 0$, this implies that $\lambda_{n+1} = \lambda_i$ for $i = 1, \ldots, n$—and from that it follows that A is a scalar multiple of the identity.

9 F Solution 9 F.

One possibility for a 7×7 matrix of the kind described is that it is skew-symmetric, in which case its determinant must be 0. (A matrix (a_{ij}) is skew-symmetric if $a_{ji} = -a_{ij}$ for all i and j.)

In the 8×8 case, there is a non-obvious but beautiful trick that solves the problem immediately: namely, form A^2 modulo 2. That is: square the matrix A, but, after doing so, reduce each entry modulo 2. What happens? A diagonal entry of A^2, say the one at position $[i, i]$ is the sum of a single 0 and (since n is even) an odd number of 1's, with signs that may be positive or negative. An off-diagonal entry of A^2, say the one at position $[i, j]$ with $i \neq j$, is the sum of two 0's and an even number of 1's, with signs that may be positive or negative. Since plus equals minus modulo 2, the signs can be ignored. Consequence: $A^2 \equiv 1$ (the identity) modulo 2. Conclusion: $\det A^2 \neq 0$, and therefore A is necessarily invertible; no additional conditions are needed.

Comment. There is, of course, nothing magical about 7 and 8; the comments about 7 remain true for $n \times n$ matrices for all odd values of n, and, similarly, the comments about 8 remain true for all even values of n.

Solution 9 G.

9 G

Yes, the converse is true, and an efficient way to prove it is via the Jordan canonical form. (Observe that $A^n \to 0$ if and only if $(T^{-1}AT)^n \to 0$.) The relevant part of Jordan theory is the assertion that (the Jordan form of) A is the direct sum of matrices of the form $\lambda+B$, where B is nilpotent (of some index k). Caution: it's possible that the numbers λ are not real. Since

$$(\lambda + B)^n = \lambda^n + \binom{n}{1}\lambda^{n-1}B + \cdots + \binom{n}{k-1}\lambda^{n-k+1}B^{k-1}$$

as soon as $n \geqq k-1$, and since $|\lambda| < 1$ implies that the coefficients tend to 0 as $n \to \infty$, it follows that if every eigenvalue of A is less than 1 in absolute value, then $A^n \to 0$.

Solution 9 H.

9 H

The sequence is always convergent; the proof is an interesting and somewhat unusual application of linear algebra.

If

$$a_{n+k} = \frac{1}{k}\sum_{j=0}^{k-1} a_{n+j},$$

then the result of applying the $k \times k$ matrix

$$A = \begin{pmatrix} 0 & 1 & 0 & \dots & 0 & 0 \\ 0 & 0 & 1 & \dots & 0 & 0 \\ 0 & 0 & 0 & \dots & 0 & 0 \\ \vdots & \vdots & \vdots & & \vdots & \vdots \\ 0 & 0 & 0 & \dots & 0 & 1 \\ \frac{1}{k} & \frac{1}{k} & \frac{1}{k} & \dots & \frac{1}{k} & \frac{1}{k} \end{pmatrix}$$

to the vector $(a_n, a_{n+1}, \dots, a_{n+k-1})$ is the next vector

$$(a_{n+1}, a_{n+2}, \dots, a_{n+k})$$

of that form. The question, therefore, is about the behavior of the powers of A.

The algebraic structure of A can be used to good effect. Consider, to begin with, the characteristic polynomial

$$\lambda^k - \frac{1}{k}\left(\lambda^{k-1} + \cdots + \lambda + 1\right).$$

Clearly 1 is a zero of this polynomial; that is 1 is an eigenvalue of A, with corresponding eigenvector $(\frac{1}{k}, \ldots, \frac{1}{k})$. Division of the characteristic polynomial by $\lambda - 1$ yields

$$\lambda^{k-1} + \frac{k-1}{k}\lambda^{k-2} + \cdots + \frac{2}{k}\lambda + \frac{1}{k}.$$

Since 1 is not a zero of the quotient, the algebraic multiplicity of 1 as an eigenvalue of A is exactly 1.

What other eigenvalues does A have? If λ is one of them, $\lambda \neq 1$, with corresponding eigenvector $x = (x_1, \ldots, x_k)$, so that $Ax = \lambda x$, then

$$\left(x_2, x_3, \ldots, x_k, \frac{1}{k}\sum_{j=1}^{k} x_j\right) = \lambda(x_1, x_2, \ldots, x_{k-1}, x_k),$$

or

$$x_2 = \lambda x_1, x_3 = \lambda x_2, \ldots, x_k = \lambda x_{k-1}, \frac{1}{k}\sum_{j=1}^{k} x_j = \lambda x_k.$$

Hence

$$\begin{aligned} x_2 = \lambda x_1, x_3 = \lambda x_2, \ldots, x_k = \lambda x_{k-1}, \left|\frac{1}{k}\sum_{j=1}^{k} x_j\right| &= |\lambda| \cdot |x_k| \\ &= |\lambda|^2 \cdot |x_{k-1}| \\ &= \cdots = |\lambda|^k \cdot |x_1|. \end{aligned}$$

Since $\lambda \neq 1$, the x's are not all equal; it follows that

$$\left|\frac{1}{k}\sum_{j=1}^{k} x_j\right|$$

is strictly less than $\max\{|xj|: j = 1, \ldots, n\}$, and hence that $|\lambda| < 1$.

The powers of a linear transformation all of whose eigenvalues are in the open unit disk converge to 0 (see 9 G). Apply that comment to the quotient transformation obtained from A by dividing out the eigenspace of the eigenvalue 1. The conclusion is that the sequence A^n converges to a matrix with one eigenvalue equal to 1 and all others equal to 0.

The limit matrix $B = (b_{ij})$, being of rank 1, is given by $b_{ij} = c_i d_j$ for some numbers c_i and d_j. The transposes $(A')^n$ converge to B', that is to the matrix whose entry in row i and column j is $c_j \cdot d_i$.

Since A leaves $(\frac{1}{k}, \ldots, \frac{1}{k})$ fixed, the same is true for B. Hence

$$\frac{1}{k} = \sum_j c_i d_j \frac{1}{k} = \frac{1}{k} c_i \sum_j d_j,$$

and therefore

$$c_1 = \cdots = c_n = \frac{1}{\sum_j d_j}.$$

Let c be the common value of these numbers.

Since A' leaves $(1, 2, \ldots, k)$ fixed, the same is true for B'. Hence

$$i = \sum_j c d_i j = c d_i \sum_j j = c d_i \frac{k(k+1)}{2},$$

and therefore

$$c d_i = \frac{i}{\frac{k(k+1)}{2}}.$$

It follows that

$$b_{ij} = \frac{j}{\frac{k(k+1)}{2}}.$$

Conclusion: the sequence $\{A^n(a_0, a_1, \ldots, a_{k-1})\}$ converges to the vector all of whose coordinates are equal to

$$\sum_{j=1}^{k} \frac{j a_{j-1}}{\frac{k(k+1)}{2}}.$$

Comment. The method generalizes to weighted means in place of averages. The result is a special case of well-known properties of Markoff matrices. Probability interpretation: start with the numbers $a_0, a_1, \ldots,$

a_{k-1} and the transition probabilities that are the entries of A. After many repetitions, the result will be a_j with probability nearly equal to

$$\frac{j}{\frac{k(k+1)}{2}}$$

(that is, the probabilities of $a_0, a_1, \ldots, a_{k-1}$ become proportional to $1, \ldots, k$).

9 I

Solution 9 I.

If A is a linear transformation on $\mathbb{C}^2$, with trace t and determinant d, then the characteristic polynomial of A is

$$\lambda^2 - t\lambda + d,$$

and therefore, by the Hamilton–Cayley equation,

$$A^2 = tA - d.$$

In principle, that's all there is to it; since this equation can be solved for A, it gives an expression (in case A is positive) for A in terms of A^2, which is in effect what is wanted. There is, however, a troublesome detail: the coefficients t and d depend on the answer (that is, A) and not on the data (that is, A^2). Once faced, the difficulty is easy to overcome: since

$$\det A^2 = (\det A)^2$$

and, for 2×2 matrices,

$$\operatorname{tr} A^2 = (\operatorname{tr} A)^2 - 2 \det A,$$

so that, for positive matrices,

$$d = \sqrt{\det A^2} \qquad \text{and} \qquad t = \sqrt{\operatorname{tr} A^2 + 2\sqrt{\det A^2}},$$

it follows that

$$A = (A^2 + \sqrt{\det A^2}) / \sqrt{\operatorname{tr} A^2 + 2\sqrt{\det A^2}},$$

To obtain the desired formula, replace A by $\sqrt{A}$. In the numerical case at hand, $\det A = 1$ and $\operatorname{tr} A = 3$, so that

$$\sqrt{A} = \frac{1}{\sqrt{5}}(a+1) = \frac{1}{\sqrt{5}}\begin{pmatrix} 3 & 1 \\ 1 & 2 \end{pmatrix}.$$

Solution 9 J.

9 J

Yes: if $\|Ux - Uy\| = \|x - y\|$ and $U(0) = 0$, then U is linear.

For the proof, observe first that since

$$\begin{aligned} \|x\|^2 - 2(x,y) + \|y\|^2 = \|x-y\|^2 &= \|Ux - Uy\|^2 \\ &= \|Ux\|^2 - 2(Ux, Uy) + \|Uy\|^2, \end{aligned}$$

it follows that

$$(x,y) = (Ux, Uy)$$

for all x and y. (Symbols such as (x,y) here refer, of course, to the inner product of the indicated vectors x and y.) Consider then for each α, β, x, and y, the difference

$$d(\alpha, \beta, x, y) = U(\alpha x + \beta y) - \alpha Ux - \beta Uy.$$

This difference is orthogonal to Uz for all z, because its inner product with Uz is equal to

$$(\alpha x + \beta y, z) - \alpha(x, z) - \beta(y, z).$$

To say that for some α, β, x, and y, the difference $d(\alpha, \beta, x, y)$ is not zero, means that there exists a non-zero vector orthogonal to the entire range of U, and hence that that range has smaller dimension than the entire space $\mathbb{V}$. If that is the case, then (induction on dimension) U is linear on $\operatorname{ran} U$; it follows, in particular, that U is bijective on $\operatorname{ran} U$. But U maps $\mathbb{V}$ onto $\operatorname{ran} U$, and that is a contradiction with the injectivity of U (which is a consequence of isometry). Conclusion: the difference under study must always be 0, which means exactly that U is linear.

Comment. An induction proof is usually thought of as starting at step number 1, but the proof just offered cavalierly ignored that step. Should the case of spaces of dimension 1 have been discussed? Yes, perhaps, and a direct proof is easy enough for them—but, easy or no, it shouldn't just be ignored. There is, however, another point of view: start at step number 0. For a vector space of dimension 0 the assertion is trivial, and the induction proof just offered reduces the statement for a space of dimension 1 to the statement for a space of dimension 0. All is well.

9 K

Solution 9 K.

The determinant of a matrix is the product of its eigenvalues (with multiplicities), and the eigenvalues of $1 + A$ are trivially calculable from those of A; the question reduces to the determination of the eigenvalues of A. If $x \perp v$, then $Ax = 0$; in other words, if $v \neq 0$, then 0 is an eigenvalue of A with corresponding eigenspace of codimension 1. (If either $u = 0$ or $v = 0$, then $A = 0$, and everything is trivial—in that case $\det(1+A) = 1$.) The eigenvalue equation $Ax = \lambda x$ is equivalent to $(x, v)u = \lambda x$. If $\lambda = 0$ (and $u \neq 0$), this implies that x is orthogonal to v; if $\lambda \neq 0$, then x is a multiple of u. Assume, with no loss, that $x = u$; then $(u, v) = \lambda$. In every case, therefore, the set of eigenvalues of A consists of 0 (with high multiplicity) together with (u, v) (with multiplicity 1 in case $(u, v) \neq 0$). The set of eigenvalues of $1 + A$, therefore, consists of 1 and $1 + (u, v)$. Conclusion: $\det(1 + A) = 1 + (u, v)$. This is negative if and only if $(u, v) < -1$.

9 L

Solution 9 L.

The product of two commuting normal transformations is normal. If, indeed, A and B are normal and $AB = BA$, then

$$A^*B = BA^*.$$

A more general statement is true: if A is normal and B is any transformation, normal or not, that commutes with A, then B commutes with A^* also. For finite-dimensional spaces that's very easy to prove: just look at a diagonal matrix in the role of A. For infinite-dimensional spaces the statement is still true, but it is not that easy to prove; it is called Fuglede's theorem. In either case, once it is known, it implies that

$$B^*A = AB^*$$

(by forming adjoints). It follows that

$$\begin{aligned}(AB)^*(AB) &= (B^*A^*)(AB) = B^*AA^*B \quad \text{(since } A \text{ is normal)}\\ &= AB^*BA^* \quad \text{(by the displayed equations above)}\\ &= (AB)(B^*A^*) \quad \text{(since } B \text{ is normal).}\end{aligned}$$

Solution 9 M.

9 M

It's good to recall that the norm of a transformation T satisfies the equation

$$\|T\|^2 = \|T^*T\|.$$

It follows, in particular, that if A and B are normal, then

$$\begin{aligned}\|AB\|^2 = \|ABB^*A^*\| &= \|AB^*BA^*\| = \|AB^*\|^2\\ &= \|BA^*\|^2 = \|BA^*AB^*\|\\ &= \|BAA^*B^*\| = \|BA\|^2.\end{aligned}$$

Conclusion: a curious tiny sliver of commutativity is always true for normal transformations.

Solution 9 N.

9 N

Yes, there exist matrices A and B such that $e^{A+B} \neq e^A e^B$, but one is not likely to stumble across them by chance. The main difficulty is that e^A does not spring to the eye; its calculation can be quite troublesome. A good trick to make the trouble go away is to work with nilpotent matrices—infinite series for them become finite. If, for instance,

$$A = \begin{pmatrix} 0 & 1 & 0 \\ 0 & 0 & 0 \\ 0 & 0 & 0 \end{pmatrix}, \qquad B = \begin{pmatrix} 0 & 0 & 0 \\ 0 & 0 & 1 \\ 0 & 0 & 0 \end{pmatrix},$$

so that

$$C = A + B = \begin{pmatrix} 0 & 1 & 0 \\ 0 & 0 & 1 \\ 0 & 0 & 0 \end{pmatrix},$$

then

$$A^2 = B^2 = C^3 = 0,$$

and, consequently,

$$e^A = 1 + A, \quad e^B = 1 + B, \quad e^C = 1 + C + \frac{C^2}{2}.$$

Consequence:

$$e^A \cdot e^B = (1 + A)(1 + B) = \begin{pmatrix} 1 & 1 & 1 \\ 0 & 1 & 1 \\ 0 & 0 & 1 \end{pmatrix},$$

whereas

$$e^C = \begin{pmatrix} 1 & 1 & \frac{1}{2} \\ 0 & 1 & 1 \\ 0 & 0 & 1 \end{pmatrix}.$$

9 O

Solution 9 O.

The answer is no; the equation $e^{A+B} = e^A e^B$ can sometimes hold for non-commuting matrices too. One way to construct an example is based on the following observations.

1. If A is a triangular matrix whose diagonal consists of distinct integral multiples of $2\pi i$, then $e^A = 1$. Reason: A is similar to a diagonal matrix.
2. If B is a diagonal matrix with distinct diagonal entries, then every matrix that commutes with B is diagonal.
3. There exist two matrices A and B such that

   a. A is triangular but not diagonal,
   b. B is diagonal,

c. each of A, B, and $A+B$ has distinct diagonal entries that are integral multiples of $2\pi i$.

Here is an example of the situation described in (3):

$$A = \begin{pmatrix} 0 & 1 \\ 0 & 2\pi i \end{pmatrix}, \qquad B = \begin{pmatrix} 2\pi i & 0 \\ 0 & -2\pi i \end{pmatrix}.$$

Conclusion from (1), (2), and (3): there exist non-commuting matrices A and B such that

$$e^A = e^B = e^{A+B} = 1.$$

Solution 9 P.

9 P

The answer is yes; for Hermitian matrices the exponential law is in fact equivalent to commutativity. Suppose, indeed, that $e^{A+B} = e^A e^B$, with A and B Hermitian; it is to be proved that A and B commute.

The Hermitian character of A and B implies that e^A and e^B are Hermitian, and, the Hermitian character of $a+b$ implies that their product is also. Consequence: e^A and e^B commute. That's a long step toward proving that A and B commute, but not quite long enough. The temptation is to take logarithms, but it's not obvious whether any such step is legitimate.

If α and β are scalars, then so are e^α and e^β, so that they commute with one another and, for that matter, with everything. From the already proved commutativity $e^A e^B = e^B e^A$, it follows that

$$e^\alpha e^\beta e^A e^B = e^\alpha e^\beta e^B e^A,$$

which can be rewritten as

$$e^{A+\alpha} e^{B+\beta} = e^{B+\beta} e^{A+\alpha}.$$

If α and β are chosen positive and sufficiently large, then the exponents in this equation become positive matrices, and the desideratum is equivalent to the statement that those exponents commute. In other words, the problem of inferring the commutativity of A and B from the commutativity of e^A and e^B has been reduced to the case in which A and B are positive matrices.

Now it's all right to take logarithms. Consider a decent subset of the positive real axis (say an interval) large enough to contain the spectra of both A and B. The logarithm function is a uniform limit of polynomials on such a set; since for a polynomial p it is true that

$$p(e^A)e^B = e^B p(e^A),$$

it follows, by passage to the limit, that

$$Ae^B = e^B A.$$

That's almost the final victory: it doesn't yet say that A commutes with B, but it says something almost that good. Apply the trick once more to infer that

$$Ap(e^B) = p(e^B)A,$$

for all polynomials, and hence, by passage to the limit once more, that

$$AB = BA,$$

and the victory is complete.

9 Q Solution 9 Q.

The answer is no; it is possible to prove that

$$0 \leqq A \leqq B$$

does not imply

$$e^A \leqq e^B$$

with only reasonable computations, which are, in fact, sort of fun.

One of the difficulties with the exponential function is that matrices A and B for which e^A and e^B are easily calculated are hard to come by. If, however, A is a 2×2 projection of rank 1, then things become easy; in that case

$$e^A = (e - 1)A + 1.$$

(It is sufficient to check this for the case in which A is diagonal; its truth in that case implies its truth in every case.)

If

$$A = \begin{pmatrix} \frac{1}{2} & \frac{1}{2} \\ \frac{1}{2} & \frac{1}{2} \end{pmatrix},$$

for which

$$b = \begin{pmatrix} x & 0 \\ 0 & y \end{pmatrix},$$

is it true that $A \leqq B$? That is: for which x and y is it true that

$$\begin{pmatrix} x & 0 \\ 0 & y \end{pmatrix} - \begin{pmatrix} \frac{1}{2} & \frac{1}{2} \\ \frac{1}{2} & \frac{1}{2} \end{pmatrix} \geqq 0\,?$$

Necessary and sufficient conditions are (just look at the diagonal entries of the difference and at its determinant):

$$(*) \qquad x - \frac{1}{2} > 0, \quad y - \frac{1}{2} > 0, \quad 2xy \geqq x + y.$$

If these conditions are satisfied, must it follow that $e^A \leqq e^B$? If that were true, then, in particular, it would be true that the top left entry of e^A would be dominated by the top left entry of e^B. The explicit expression for e^A shows that the top left entry in it is

$$(e-1)\frac{1}{2} + 1 = \frac{e+1}{2}.$$

The top left entry in e^B is e^x. The question, therefore, is this: do there exist x and y satisfying the inequalities $(*)$ such that $e^x < \frac{e+1}{2}$? The answer is yes, easily.

The inequality $e^x < \frac{e+1}{2}$ says (or so my hand calculator assures me) the same thing as $x < \log 1.85914\ldots$ (natural log of course), or

$$x < .620115\ldots .$$

If, for instance, $x = .6$, can y be found so that $(*)$ is satisfied? The only condition left to satisfy is $2xy \geqq x + y$; with $x = .6$ that becomes $1.2y \geqq y + .6$, which is equivalent to $y \geqq 3$. Conclusion: if

$$A = \begin{pmatrix} \frac{1}{2} & \frac{1}{2} \\ \frac{1}{2} & \frac{1}{2} \end{pmatrix} \quad \text{and} \quad B = \begin{pmatrix} .6 & 0 \\ 0 & 3 \end{pmatrix},$$

then $0 \leqq A \leqq B$ is true but $e^A \leqq e^B$ is false.

Chapter 10. Algebra

10 A

Solution 10 A.

The answer is no; $\mathbb{R}$ mod 1 is not the additive group of a ring with unity. Suppose it were, and suppose that e is the unity. Since in the group $\mathbb{R}$ every element has a "half", that is, for every u there exists a v such that $v + v = u$, it is true in particular that there exists a number x such that $x + x = e$. If, however, $y = \frac{1}{2}$, then $y + y = 0$, and since e is the unity, it must be true that $ey = y$—but there is trouble, because

$$ey = (x + x)y = xy + xy = x(y + y) = x0 = 0.$$

This is a contradiction, and the proof is complete.

Comment. What happens if the group $\mathbb{R}$ is replaced by the group $\mathbb{Q}$ of rational numbers ?

10 B

Solution 10 B.

1. No, the additive group of $\mathbb{Q}$ is not isomorphic to $\mathbb{Q}^{*^+}$). One way to see that is to observe that $\mathbb{Q}$ is divisible, and in particular divisible by 2, where $\mathbb{Q}^{*^+}$ is not. To say that $\mathbb{Q}$ is divisible means that for every x in $\mathbb{Q}$ and for every positive integer n there exists an element y in $\mathbb{Q}$ such that $ny = x$. (Is the meaning of the symbol "ny" clear? It is intended to denote the sum of n copies of y.) Divisibility by 2, means, of course, that for every x in $\mathbb{Q}$ there exists an element y in $\mathbb{Q}$ such that $y + y = x$. If the group $\mathbb{Q}$ were isomorphic to $\mathbb{Q}^{*^+}$, it would follow that for every x in $\mathbb{Q}^{*^+}$ there exists an element y such that $yy = x$, that is that every x in $\mathbb{Q}^{*^+}$ has a square root in $\mathbb{Q}$—which is a famous falsehood. (Remember the lore about $\sqrt{2}$.)

2. No, the multiplicative groups $\mathbb{R}^*$ and $\mathbb{C}^*$ are not isomorphic. One way to see that is to look, in both groups, for solutions of the equation $x^3 = 1$: in the group $\mathbb{R}^*$ there can be only one such x, but in $\mathbb{C}^*$ there are three.

3. No, there is no field whose additive group is isomorphic to its multiplicative group. To see that, observe, to begin with, that such a field could not be finite (because the multiplicative group has one element less than the additive group). In the infinite case, if there were such an isomorphism, it would map elements of multiplicative order 2 to elements of additive order 2. The number of solutions of the equation $x^2 = 1$ in an infinite field is either one (characteristic 2) or two (characteristic different from 2), whereas the number of solutions of the equation $2x = 0$ is either infinite (characteristic 2) or one (characteristic different from 2).

Solution 10 C.

10 C

Yes, if all the elementary symmetric functions of a finite set of real numbers $a_1, a_2, \ldots, a_n$ are positive, then all a's must be positive. An easy way to prove the assertion is to consider the polynomial

$$(x + a_1)(x + a_2) \cdots (x + a_n).$$

Multiply it out and look, and find that the leading coefficient is 1, and the other coefficients are exactly the elementary symmetric functions in the a's. Under the present assumptions all those coefficients are positive, and, therefore, the zeroes of the polynomial must all be negative. Those zeroes, however, can be read off from the product form: they are exactly the numbers $-a_1, -a_2, \ldots, -a_n$, and that completes the proof.

Solution 10 D.

10 D

Yes, it's true that if a polynomial $p(x, y)$, with coefficients in any ring $\mathbb{R}$, vanishes when $x = y$, then it is divisible by $x - y$. Consider, indeed, the polynomial p^- in the single variable x, with coefficients in the polynomial ring $\mathbb{R}[y]$ defined by

$$p^-(x) = p(x, y).$$

Since the division algorithm works for $\mathbb{R}[y]$, and since, by assumption, y is a zero of $p^-(x)$ in that ring, it follows that $p_-(x)$ is divisible by $x - y$, and hence that $p(x, y)$ is divisible by $x - y$.

Comment. The reasoning has a slippery point: the vanishing of $p^{-}(x)$ must be interpreted in the formal sense, not in the functional sense. Example: if $\mathbb{R}$ is the ring of integers modulo 2, and if

$$p(x, y) = xy + x,$$

then $p(x, x) = 0$ when $x = 0$ and when $x = 1$ (so that $p(x, x)$ vanishes in the functional sense), but $p(x, x)$ $(= x^2 + x)$ is not formally 0, and, of course, $p(x, y)$ is not divisible by $x - y$.

10 E

Solution 10 E.

Yes, such examples do exist, but some sophisticated methods are needed to produce them. Here is one way to proceed.

Consider the field $\mathbb{R}(x)$ of all rational functions in one indeterminate, over the field $\mathbb{R}$ of real numbers, and the field $\mathbb{R}(x, y)$ of all rational functions in two indeterminates over $\mathbb{R}$. In both cases the additive group is a vector space over the field $\mathbb{Q}$ of rational numbers with a Hamel basis of the power of the continuum. In both cases the multiplicative group is the direct product of $\mathbb{R}^*$ and the free abelian group on continuum many generators (namely the irreducible polynomials). The two fields are, however, not isomorphic, because they have transcendence bases consisting of different numbers of elements (namely one for $\mathbb{R}(x)$ and two for $\mathbb{R}(x, y)$).

10 F

Solution 10 F.

The answer is no: if two different monic polynomials attain two distinct values at the same places, then they are identical.

Suppose that p and q are polynomials and a and b are two distinct numbers such that

$$\text{null}\,(p - a) = \text{null}\,(q - a)$$

and

$$\text{null}\,(p - b) = \text{null}\,(q - b)$$

(where "null" is an ad hoc symbol to indicate the set of zeroes of a polynomial). Choose the notation so that $\deg p \geqq \deg q$. If

$$(p(x) - a)(q(x) - b) = 0$$

for some x, then either $p(x) = a$, in which case $q(x) = a$ also, and therefore $p(x) - q(x) = 0$, or $q(x) = b$, in which case $p(x) - q(x) = 0$ again. Consequence:

$$\text{null } (p - a)(q - b) \subset \text{ null } (p - q).$$

If $x \in$ null $(p - a)(p - b)$, then either

$$x \in \text{ null } (p - a),$$

or

$$x \in \text{ null } (p - b),$$

but not both (since $a \neq b$). Suppose that x is a zero of $p - a$ of order k; then it is a zero of the derivative $(p - a)'$ of order $k - 1$. Since it is a zero of $p - q$ of order at least 1, it follows that it is a zero of $p'(p - q)$ of order at least k. Consequence: $(p - a)(p - b)$ divides $p'(p - q)$.

If $\deg p = n$, then $\deg (p - a)(p - b) = 2n$, and

$$\deg p'(p - q) \leqq (n - 1) + n = 2n - 1.$$

Consequence: $p'(p - q) = 0$. If p' is identically 0, then p is a constant, and therefore q is a constant; in this case the desired equality is trivially true. If p' is not identically zero, then $p - q$ is, and the conclusion holds again.

Solution 10 G.

10 G

The answer is no: if an element in a ring with unity has a left inverse but no right inverse, then it has infinitely many left inverses. The proof imitates the U, V example.

Suppose that x has a left inverse y, and consider the elements

$$y + x^n(1 - xy), \qquad n = 0, 1, 2, \ldots .$$

Assertion: they are all left inverses of x and no two of them are equal.

Since

$$(y + x^n(1 - xy))x = yx + x^{n+1} - x^{n+1} = 1,$$

they are indeed left inverses.

If $y + x^n(1 - xy) = y + x^m(1 - xy)$, or, equivalently, if

$$x^n(1 - xy) = x^m(1 - xy),$$

and if, say, $n > m$, multiply by y^n on the left to get

$$1 - xy = y^n x^m(1 - xy).$$

Since $y^n x^m = y^{n-m} y^m x^m = y^{n-m}$, it follows that if $k = n - m$, then $k \geqq 1$ and

$$\begin{aligned} 1 - xy = y^k(1 - xy) &= y^k - y^k xy \\ &= y^k - y^{k-1}(yx)y = y^k - y^{k-1}y \\ &= y^k - y^k = 0, \end{aligned}$$

and that contradicts the assumption that x has no right inverse.

10 H

Solution 10 H.

The answer is no; the product of the elements of some block in the sequence must always be the identity. For the proof, form the n possible partial products $\prod_{i=1}^{q} a_i$. If they are all distinct, then they are the n different elements of the group, and therefore one of them is 1 (the identity element), and the problem is solved. If, however, two of them coincide, say, for instance

$$\prod_{i=1}^{10} a_i = \prod_{i=1}^{20} a_i$$

then multiply both sides of the equation by the inverse of the left side and get

$$1 = \prod_{i=11}^{20} a_i.$$

Solution 10 I.

10 I

The best way to attack the question is to ask a more special question first: if K is a subgroup of an additive abelian group G, which subsets of G are cosets of K?

To get a grip on the answer, consider a coset $S = a + K$ of K, and note that

$$S - S = (a + K) - (a + K) = K - K \subset K$$

(because K is a subgroup), and

$$S + K = a + K + K \subset a + K = S.$$

These two necessary conditions together are, in fact, sufficient as well: if $S - S \subset K$ and $S + K \subset S$, then S is a coset of K. To prove that, choose an arbitrary element a of S and note that

$$S \subset S + (a - S) = a + (S - S) \subset a + K \subset S + K \subset S.$$

It follows that equality holds throughout, and therefore that $S = a + K$. Incidentally: the proof shows that $S - S = K$—the coset S is eager to tell us which subgroup it is a coset of—and the proof shows that $S + K = S$.

Suppose now that all that is known is that S is a subset of some subgroup K of G, but K is not known. Whatever K is, what is known is that $S + S - S \subset S + K \subset S$ (by the preceding result), and that inclusion does the work that needs to be done. That is: if S is a subset of G such that $S + S - S \subset S$, put $K = S - S$ (if there is a subgroup K, that's what it must be), and calculate with K. Since

$$K - K = (S - S) - (S - S) = (S + S - S) - S \subset S - S = K,$$

it follows that K is indeed a subgroup, and, moreover, a subgroup such that $S - S \subset K$ (in fact $S - S = K$) and $S + K \subset S$, and the desired conclusion follows from the paragraph above. Explicitly: S is a coset if and only if $S + S - S \subset S$.

Solution 10 J.

10 J

The only automorphism that there is room for in a group of order 2 is the trivial one—the pleasant result is that every group with more than two elements has a non-trivial automorphism.

The proof is easy, but not totally elementary. If the group is not abelian, choose an element a, say, not in the center, and consider the inner automorphism it induces—that is the mapping that sends each x to $a^{-1}xa$. If the group is abelian, but not every element is of order 2, consider the mapping (automorphism) that sends every element to its own inverse. If, finally, every element is of order 2, then the group is a vector space of dimension at least 2 over the field of two elements, and, as such it has a basis; the interchange of two elements of a basis is an automorphism.

10 K

Solution 10 K.

Yes, there is at least one finite group with an automorphism that maps exactly three fourths of the elements onto their own inverses. One example is the group of all rigid motions of the square, or, in other words, if the square is thought of as the pattern

$$\begin{matrix} 3 & 2 \\ 4 & 1, \end{matrix}$$

then the permutation group generated by the rotation $(1\ 2\ 3\ 4)$ and the reflection $(2\ 4)$. The order of the group is 8, and straightforward verification shows that the inner automorphism induced by $(2\ 4)$ interchanges $(1\ 2)(3\ 4)$ and $(1\ 4)(2\ 3)$ and maps each of the other six elements onto its own inverse.

Comment. If an automorphism A maps more than half of the elements of a finite group onto their own inverses, then A^2 is the identity automorphism. Indeed: if $Ax = x^{-1}$, then

$$A^2x = A(Ax) = Ax^{-1} = (Ax)^{-1} = (x^{-1})^{-1} = x,$$

so that

$$\{x\colon Ax = x^{-1}\} \subset \{x\colon A^2x = x\}.$$

The latter set is a subgroup, and a subgroup that contains more than half the elements of a group is the whole group.

Solution 10 L.

10 L

No, there is no finite group with an automorphism that maps exactly four fifths of the elements onto their own inverses—in fact if A is an automorphism of a finite group G such that

$$\mu(\{x: Ax = x^{-1}\}) > \frac{3}{4}\mu(G),$$

where μ stands for "the number of elements of", then $Ax = x^{-1}$ for every x in G (and G is necessarily abelian).

For the proof, write $E = \{x: Ax = x^{-1}\}$ and let x_0 be a (temporarily) fixed element of E. Since $\mu(x_0E) = \mu(E) > \frac{3}{4}\mu(G)$, it follows that $\mu(E \cap x_0E) > \frac{1}{2}\mu(G)$. If $x \in E \cap x_0E$, then $x \in E$ and $x = x_0y$ for some y in E. It follows that $Ay = y^{-1}$ and

$$A(x_0y) = Ax = x^{-1} = (x_0y)^{-1} = y^{-1}x_0^{-1}.$$

Since, however

$$A(x_0y) = A(x_0)A(y) = x_0^{-1}y^{-1},$$

it follows that y commutes with x_0, and therefore so does x. In other words, $E \cap x_0E$ is included in the centralizer of x_0. Since that centralizer is a subgroup of G and the number of its elements is greater than $\frac{1}{2}\mu(G)$, it must coincide with G. (Contemplate its index.) Consequence: x_0 is in the center of G. Since this is true for all x_0 in E, the set E is included in the center of G. It follows that if x and y are in E, then

$$A(xy) = A(yx) = (Ay)(Ax) = y^{-1}x^{-1} = (xy)^{-1},$$

so that $xy \in E$. This, in turn, implies that E is a subgroup of G and hence (counting argument as before) coincides with G.

Comment. Both the statement and the proof remain unchanged if "finite" is replaced by "compact" and μ is interpreted not as number but as (Haar) measure.

Solution 10 M.

10 M

No, not every semigroup with cancellation can be embedded in a group. Here is a nasty example.

Let S be the set of all (possibly empty) words (finite words) formable with the eight letters a, b, c, d, x, y, u, v; let multiplication be defined as juxtaposition. That is: the product of, say, abc with $axayau$ (in that order) is $abcaxayau$; the product of b with b is bb; the product of $vuyx$ with $\varnothing$ is $vuyx$. With that definition S is a semigroup with cancellation (both cancellation laws hold); it is sometimes called the free semigroup generated by the given eight symbols.

Use S to manufacture a new semigroup by making the identifications

$$ax = by,$$

$$cx = dy,$$

$$au = bv,$$

but NOT the identification $cu = dv$. The word "identification" is common in contexts such as this, but it is, to tell the truth, slightly sloppy—it needs to be understood precisely before it can be used. In the present context, which is about as simple as they ever come, to identify means to define the equivalence relation indicated by the specified equations, and to interpret multiplication so as to harmonize with that relation. In more detail, define two words to be equivalent, indicated by, say, the symbol $\equiv$, if one can be obtained from the other by the application of a (finite) number of substitutions of the forms

$$ax \to by, \qquad by \to ax,$$

$$cx \to dy, \qquad dy \to cx,$$

$$au \to bv, \qquad bv \to au.$$

Examples:

$$axax \equiv byax \equiv axby \equiv byby,$$

and

$$auuadydy \equiv bvuacxdy.$$

It is easy enough to check that the relation is indeed an equivalence (reflexive, symmetric, and transitive)—all it takes is to remember the definitions.

Once the equivalence relation is at hand, there are two popular ways of using it to construct a new semigroup. One is to define the product of two equivalence classes by selecting a word from each, multiplying the two words, and forming the equivalence class of the result. This approach needs to be justified by the insertion of a lemma: it must be proved that if the selected words are replaced by equivalent ones, the resulting equivalence class is not changed. The other approach produces a semigroup not by manufacturing new kinds of elements (equivalence classes), but by reinterpreting the meaning of equality: the elements in the second approach are the same as they were before (words), but two words are regarded as equal if and only if they are equivalent. The second approach is the one most of us usually adopt when we think about modular arithmetic: the sum of 4 and 6 is 10, and the sum of 4 and 6 modulo 7 is also 10, but we allow ourselves to think of 10 as being the "same" as 3.

The new semigroup, call it T, is a semigroup with unit and with cancellation, and, in it the elements cu and dv are different. Assertion: the semigroup T cannot be embedded into a group. Indeed, if that were possible, then it would follow (in such a group) that

$$cu = c(a^{-1}bv) = c(a^{-1}by)y^{-1}v = (cxy^{-1})v = dv,$$

a contradiction.

Comment. A commutative semigroup with cancellation is always embeddable in a group by the usual process of manufacturing the integers from the natural numbers.

Solution 10 N.

10 N

The answer is no: a group cannot be the union of two proper subgroups. Suppose, indeed, that G is a group and that H and K are subgroups such that $G = H \cup K$, with $H \neq G$ and $K \neq G$. It follows that the set-theoretic differences $H \setminus K$ and $K \setminus H$ cannot be empty. (Some people always use $H \setminus K$ for set-theoretic difference and others use $H - K$. In contexts such as the present one there is a danger of misunderstanding: $H - K$ might stand for the group-theoretic difference, that occurred in 10 I, instead of the set-theoretic one.) Consequence: there exist elements x and y with x in $H \setminus K$ and y in $K \setminus H$. Put $z = xy$, and ask: where can z be? If $z \in H$, then, since $y = x^{-1}z$, it follows that $y \in H$ (which is false), and if $z \in K$, then, since $x = zy^{-1}$, it follows that $x \in K$ (which is false). A contradiction has arrived: z cannot be in either H or K.

Comment. Compare this problem with 9 A, the similar one about vector spaces. The result was that a real vector space could not be the union of a finite number of proper subspaces; here the result is the more modest one about only two proper subspaces. That's in the nature of things: groups can be finite, and finite groups are quite likely to be the unions of finitely many proper subgroups. For one example, consider any group that is not cyclic, and observe that it is the union of all of its cyclic subgroups.

10 O

Solution 10 O.

The answer is yes even to the stronger question: yes, there exist abelian groups that have no maximal subgroups.

An interesting way to approach the proof is to show that there exist infinite abelian groups such that all their proper subgroups are finite. Indeed: suppose that G is a group with those properties and that H is a proper subgroup of G, so that H is finite. Consider an element x of G that does not belong to H. Assertion: the cyclic subgroup generated by x must be finite (for otherwise the cyclic subgroup generated by x^2 would be an infinite proper subgroup), and it follows from the abelian character of G that the group generated by H and x together is finite and hence is a proper subgroup different from H that includes H.

The problem has therefore been reduced to exhibiting an infinite abelian group all of whose proper subgroups are finite. That's not hard: one concrete example is the additive group G of all dyadic rational numbers modulo 1. In other words: the elements of G are the numbers of the form $\frac{m}{2^n}$ where m is an arbitrary integer and $n = 0, 1, 2, \ldots$; addition in G is defined to be ordinary numerical addition followed by reduction modulo 1. The meaning of the latter phrase can be perfectly explained by a single example: the ordinary sum of $\frac{3}{4}$ and $\frac{5}{8}$ is $\frac{11}{8}$, which reduces to $\frac{3}{8}$ when reduced modulo 1. (An isomorphic description of G is as the set of all complex roots of unity whose order is a power of 2.) What must be proved is that if H is a proper subgroup of G, then H is finite.

Suppose that $\frac{m}{2^n}$ is in H; there is no loss of generality in assuming that m is odd. Since m and 2^n are relatively prime, there exist integers a and b such that $am + b2^n = 1$. Consequence:

$$\frac{1}{2^n} = \frac{am + b2^n}{2^n} = \frac{am}{2^n} + b;$$

since $\frac{am}{2^n} \in H$ (and $b \equiv 0 \bmod 1$), it follows that $\frac{1}{2^n} \in H$.

Since H is proper, either H is the trivial subgroup $\{0\}$, or there exists a largest integer n_0 such that $\frac{1}{2^{n_0}} \in H$; in that case H consists exactly of the multiples of $\frac{1}{2^{n_0}}$. Reason: clearly all those multiples belong to H; if, conversely, $\frac{m}{2^n} \in H$ (with an odd m), then, by the preceding paragraph, $n \leqq n_0$.

Solution 10 P.

10 P

The answer is no: there exist two non-isomorphic groups each of which is isomorphic to a subgroup of the other. Here is a possible example. Let $\mathbb{Q}$ be the additive group of rational numbers, let $\mathbb{Z}$ be the additive group of integers, and define groups $\mathbb{G}$ and $\mathbb{H}$ as follows: $\mathbb{G}$ is the direct product of infinitely many copies of $\mathbb{Q}$,

$$\mathbb{G} = \mathbb{Q} \otimes \mathbb{Q} \otimes \mathbb{Q} \otimes \cdots,$$

and $\mathbb{H}$ is the direct product of $\mathbb{Z}$ with $\mathbb{G}$,

$$\mathbb{H} = \mathbb{Z} \otimes \mathbb{G} = \mathbb{Z} \otimes \mathbb{Q} \otimes \mathbb{Q} \otimes \mathbb{Q} \otimes \cdots.$$

Another way of saying the same thing is this: $\mathbb{G}$ is the set of all functions, defined on an infinite domain, with values in $\mathbb{Q}$, and $\mathbb{H}$ is the set of all ordered pairs (n, x) with $n \in \mathbb{Z}$ and $x \in \mathbb{G}$. Obviously $\mathbb{G}$ is (isomorphic to) a subgroup of $\mathbb{H}$. Since the infinite domain can be put in one-to-one correspondence with the same domain with one point removed, it follows also that $\mathbb{H}$ is isomorphic to a subgroup of G. To see that $\mathbb{G}$ and $\mathbb{H}$ are not isomorphic, note that $\mathbb{G}$ is divisible and $\mathbb{H}$ is not.

Chapter 11. Sets

11 A

Solution 11 A.

The answer is no.

A line in the plane has measure zero (planar measure, that is), and therefore so does the union of countably many lines.

A line is nowhere dense in the plane, and therefore a countable union of lines is a set of the first category (in the sense of Baire); since the plane is a complete metric space, such a countable union cannot exhaust the plane.

Given countably many lines, consider a line distinct from all of them. (Such a line must exist, just because there are more than countably many lines.) The new line intersects each of the given ones in countably many points, and the set of all those intersections constitutes a countable set; since each line (and, in particular, the new line) is uncountable, a countable set cannot exhaust it. Alternatively: consider an arbitrary non-degenerate circle, and count its intersections with each of the given lines.

The clue is size: any measure of size, such as measure, category, cardinal number, or dimension (use the so-called F_σ theorem in dimension theory) yields a proof that the plane is not the union of countably many lines.

11 B

Solution 11 B.

The answer, perhaps somewhat surprisingly, is no: a countably infinite set can have an uncountable collection of non-empty subsets any two of which are almost disjoint.

Here is one way to do it. Assign to each real number x a countable set Q_x of rational numbers that has x as its only cluster point (by, for instance, finding a sequence of rational numbers below x that converges to x). The Q_x's are all subsets of the countable set $\mathbb{Q}$ of rational numbers, and there are uncountably many of them. The intersection of any two distinct ones must be finite—for if $Q_x \cap Q_y$ is infinite, then the unique cluster points of Q_x and Q_y must coincide.

An enlightening and different solution is this. Assign to each angle θ, $0 \leqq \theta < 2\pi$, the set of those lattice points (in the plane) that are inside a band of width 2 and angle of inclination θ. (A "band" here is the closed set between two parallel lines.) Every such band contains at least one lattice point, because no point in the plane is farther than $\frac{\sqrt{2}}{2}$ from a lattice point. The intersection of two bands is a bounded set, and, therefore, can contain only finitely many lattice points.

Solution 11 C.

11 C

The answer this time is not the surprising one, but to find a proof might take a little time.

The assertion is that if a collection $\mathfrak{C}$ of (non-empty) sets of positive integers has the property that the intersection of any two of them has not more than 1000 elements, then the collection is countable. For the proof, for each set E in $\mathfrak{C}$ let $\Phi(E)$ be the set of the 1001 smallest elements in E (and let $\Phi(E)$ be E in case the number of elements in E is less than 1001). The mapping Φ has $\mathfrak{C}$ for its domain; its range is included in $(\mathbb{Z}^+)^{1001}$—a countable set. Assertion: the mapping Φ is one-to-one. Reason: if $\Phi(E) = \Phi(F)$, then E and F have more than 1000 elements in common, and therefore, by the assumption on $\mathfrak{C}$, they are identical. The proof is over: the one-to-one character of Φ implies that $\mathfrak{C}$ is as countable as the range of Φ.

Solution 11 D.

11 D

Much to most people's astonishment, the answer is yes—$\mathbb{R}$ can be made a vector space over $\mathbb{C}$. The reason belongs to what might be called the pathology of set theory: while $\mathbb{R}$ and $\mathbb{C}$ are intuitively and geometrically different, set theoretically and additively (!) they are the same. That is: the additive group $\mathbb{R}$ is isomorphic to the additive group $\mathbb{C}$. The point is that both $\mathbb{R}$ and $\mathbb{C}$ are vector spaces over the field $\mathbb{Q}$ of rational numbers (that's obvious), and, what's more, they are vector spaces of the same dimension (that just requires a look at the cardinal numbers involved). Consequence: $\mathbb{R}$ and $\mathbb{C}$ are isomorphic as vector spaces over $\mathbb{Q}$, and hence as additive groups. Once that is realized, the original question becomes nothing more than this: can $\mathbb{C}$ be made a vector space over $\mathbb{C}$? Since the answer to that is trivially yes, the answer to the original question is equally yes.

Comment. For ultra-sophisticated algebraists, the question should have been phrased this way: is there an injection from $\mathbb{C}$ into the ring of endomorphisms of $\mathbb{R}$? The answer then is obviously yes, since the additive group $\mathbb{R}$ is isomorphic to the additive group $\mathbb{C}$, and hence (via $\mathbb{C}$ acting on itself) there is a natural injection from $\mathbb{C}$ into the ring of endomorphisms of $\mathbb{R}$.

The question originally arose as the result of a student misunderstanding of the classical easy question: the student is the one who, by mistake, interchanged $\mathbb{R}$ and $\mathbb{C}$, and was then horrified by the result.

11 E Solution 11 E.

The answer is no. Suppose indeed that f is a function, $f: \mathbb{R} \to \mathbb{R}$, of the kind asked about. The set $f(\mathbb{Q})$ is countable, because $\mathbb{Q}$ is, and the set $f(\mathbb{R}-\mathbb{Q})$ is countable, because, by hypothesis, it is included in $\mathbb{Q}$. Consequence: the range of f is countable. By the intermediate value theorem, the range of a continuous function that takes at least two distinct values must be uncountable—a contradiction has arrived.

11 F Solution 11 F.

The answer is yes, but the reason is "pathological"—compare Solution 11 D. The point is that the same question for $\mathbb{R}^2$ in place of $\mathbb{R}$ is obviously yes—just use the four quadrants, with the boundaries included or excluded at will, and with the origin put into any one of them at will. (For example: count the strictly positive x and y axes as parts of the first quadrant, the strictly negative axes as parts of the third, and leave the second and fourth quadrants open. The origin can be put into any of the four sets so described.) Once the answer is known for $\mathbb{R}^2$, then, in view of the fact that the question is purely group theoretic, and in view of the fact that as additive groups $\mathbb{R}^2$ and $\mathbb{R}$ are isomorphic, the result follows for $\mathbb{R}$ also. Why are the additive groups $\mathbb{R}^2$ and $\mathbb{R}$ isomorphic? Think about Hamel bases.

Comment. The method also solves the question about three uncountable semigroups: just use three sectors issuing from the origin in $\mathbb{R}^2$ (and then their isomorphic versions in $\mathbb{R}$). For example, the three sets could be the first quadrant, the second quadrant, and the lower half plane, with some reasonable (but arbitrary) decisions about the boundaries. The method works, in fact, for any pertinent cardinal number in place of 4, including in particular the numbers $\aleph_0$ and $2^{\aleph_0}$.

The question about "anti-semigroups" (sets totally not closed under addition and multiplication) can stump someone only in this context —observe, for instance, that a possible answer to it is the closed interval $[10, 11]$.

Solution 11 G.

11 G

The answer is probably not the most obviously expected number; it is 14. The way to get it is kind of fun too. Two things have to be proved: that no set yields more than 14, and that some set does yield that many.

For any set A, write A^+ as an abbreviation for A'^-. Since

$$A' \subset A^+,$$

it follows that

$$A^{+\prime} \subset A,$$

and hence that

$$A^{++} \subset A^-.$$

If A is closed (so that $A^- = A$), then $A^{++} \subset A$. An application of this result to the (closed) set A^+ yields

$$A^{+++} \subset A^+.$$

If, on the other hand, $A \subset B$, then $B' \subset A'$, and therefore

$$B^+ \subset A^+.$$

It follows that if A is closed, so that $A^{++} \subset A$, then

$$A^+ \subset A^{+++},$$

and therefore, if A is closed, then

$$A^+ = A^{+++}.$$

Written out in detail, this becomes the identity

$$A^{-\prime-} = A^{-\prime-\prime-\prime-}.$$

This identity implies that the fourth term of the sequence

$$A, A^{-}, A^{-\prime}, A^{-\prime-}, A^{-\prime-\prime}, \ldots$$

is the same as its eighth term, and that, therefore, the sequence cannot contain more than seven distinct terms. Since the same result applies to A' (that is, the sequence

$$A', A'^{-}, A'^{-\prime}, A'^{-\prime-}, A'^{-\prime-\prime}, \ldots$$

cannot contain more than seven distinct terms), it follows that 14 is indeed an upper bound.

It remains to exhibit a set for which 14 is attained. That, it turns out, can be done in the line; one example for which the upper bound is attained is

$$A = \left(0, \frac{1}{3}\right) \cup \left(\mathbb{Q} \cap \left(\frac{1}{3}, \frac{2}{3}\right)\right) \cup \{1\}.$$

For "computational" purposes it is convenient to indicate this set as follows:

$$[0] \quad (+++) \quad [0] \quad (\#\#\#) \quad [0] \quad (000) \quad [+].$$

The meaning of the notation is as follows. A single $[+]$ or a single $[0]$ indicate, respectively, a point that is present or absent from the set; a triple $(+++)$ or a triple (000) indicate, respectively, an open interval that is present or absent from the set; and a triple $(\#\#\#)$ indicates a set dense in a subinterval with the property that its complement is also dense in that subinterval.

If we start with A and form, one after another, its closure, complement, closure, etc., we get

[0]	$(+++)$	[0]	(###)	[0]	(000)	[+]
[+]	$(+++)$	[+]	$(+++)$	[+]	(000)	[+]
[0]	(000)	[0]	(000)	[0]	$[+++]$	[0]
[0]	(000)	[0]	(000)	[+]	$[+++]$	[+]
[+]	$(+++)$	[+]	$(+++)$	[0]	(000)	[0]
[+]	$(+++)$	[+]	$(+++)$	[+]	(000)	[0]
[0]	(000)	[0]	(000)	[0]	$(+++)$	[+];

if we start with A' and proceed similarly, we get

[+]	(000)	[+]	(###)	[+]	$(+++)$	[0]
[+]	(000)	[+]	$(+++)$	[+]	$(+++)$	[+]
[0]	$(+++)$	[0]	(000)	[0]	(000)	[0]
[+]	$(+++)$	[+]	(000)	[0]	(000)	[0]
[0]	(000)	[0]	$(+++)$	[+]	$(+++)$	[+]
[0]	(000)	[+]	$(+++)$	[+]	$(+++)$	[+]
[0]	$(+++)$	[0]	(000)	[0]	(000)	[0].

That's it; that settles everything.

Comment. The plane, about which the question was originally asked, has nothing to do with the matter. The question can be asked about any topological space, and the proof that 14 is an upper bound works for every space.

Solution 11 H.

11 H

No, not necessarily; there exist two closed convex sets in the plane whose sum is not closed. Example: let A be the convex hull of the closed upper right half of a hyperbola, that is, the region

$$A = \{(x, y) \colon x > 0, y > 0, xy \geqq 1\},$$

and let B be the x-axis,

$$B = \{(x, y) \colon y = 0\}.$$

The sum $A+B$ is the open upper half plane,

$$A+B=\{(x,y):y>0\}.$$

11 I

Solution 11 I.

The answer is 2: the sum $C+C$ is the closed interval $[0,2]$.

One useful way to think of the Cantor set is as the set of all those numbers in $[0,1]$ that are representable in the ternary system by 0's and 2's only. (Example: $\frac{1}{3}$. To be sure $\frac{1}{3} = .1000\ldots$, but $\frac{1}{3}$ is also representable as $.0222\ldots$.) It follows that the set $\frac{1}{2}C$ is the set of all those numbers in $[0,1]$ that are representable in the ternary system by 0's and 1's only. Assertion: $\frac{1}{2}C+\frac{1}{2}C=[0,1]$.

For the proof, consider the ternary representation of an arbitrary x in $[0,1]$, and define y and z as follows:

> to get y, use the digit 0 in the same places where it occurs in x, and use the digit 1 whenever the corresponding digit in x is either 1 or 2;
>
> to get z, use the digit 0 whenever the corresponding digit in x is either 0 or 1, and use the digit 1 whenever the corresponding digit in x is 2.

In other words: give all the 1's in x to y, and split all the 2's in x half-and-half between y and z. Example: if

$$x=.012012012\ldots,$$

then

$$y=.011011011\ldots,$$

and

$$z=.001001001\ldots.$$

Consequence: y and z are in $\frac{1}{2}C$ and $y+z=x$. Conclusion, as promised: $\frac{1}{2}C+\frac{1}{2}C=[0,1]$, so that $C+C=[0,2]$.

Solution 11 J.

11 J

It might appear that removing the restriction on diameters can only enlarge the intersection, but it turns out that that is not so; the intersection can be empty. The motivation for the construction is the prototypical example of a decreasing sequence with empty intersection—it depends on the possibility of metrizing the set of positive integers so that its tails become closed balls. Once the idea is conceived, it is not so difficult to carry it out: define the distance between the positive integers n and m to be

$$d(n, m) = 1 + \frac{1}{n+m}$$

whenever $n \neq m$. (When $n = m$, the distance must be 0, of course.) With this definition, the tail $I_n = \{n, n+1, n+2, \ldots\}$ becomes a closed ball; in fact I_n is the closed ball with center n and radius $1 + \frac{1}{2}n$. Indeed: for which positive integers x does it happen that

$$1 + \frac{1}{n+x} \leqq 1 + \frac{1}{2n}?$$

The answer is obtained by solving this inequality, and that's a trivial problem; the inequality holds if and only if $x \geqq n$.

Comment. The conscientious reader will, of course, worry about whether the metric space here described is really complete. The worry will go away with the realization that in this space an open ball with center n_0 and radius less than 1 contains nothing except n_0. (In other words, every singleton is an open set; topologically speaking the space is discrete.) It follows that the only Cauchy sequences are ultimately constant, and, therefore, that the space is trivially complete.

Solution 11 K.

11 K

The answer is no: it is possible for a decreasing sequence of non-empty, closed, bounded, and convex sets to have an empty intersection. One example can be seen in the space $\mathbb{C}[0, 1]$ of continuous functions on the unit interval. Indeed, let E_n be the set of all those continuous functions on $[0, 1]$ that map $[0, 1]$ into itself, take the value 0 at 0, and take the value 1 throughout the interval $[\frac{1}{n}, 1]$. Clearly E_n is non-empty, closed, bounded, and convex, and the sequence $\{E_n\}$ is decreasing—but, equally clearly, no continuous function can belong to all the E_n's.

Comment. The expert might be interested to know that such a counterexample cannot exist in Hilbert space. Reason: for convex sets the strong topology coincides with the weak topology and the non-emptiness conclusion follows immediately from the weak compactness of the sets in question.

Chapter 12. Spaces

12 A

Solution 12 A.

The answer is a non-surprising affirmative, but it might take more than a few seconds to think of a proof. Here is one possibility.

If C is a countable subset of $\mathbb{R}^2$, and if P and Q are in the complement $\mathbb{R}^2 - C$, join P and Q by a segment and form the perpendicular bisector of that segment (see Figure 80). For each point X on that bisector, form the polygonal line $PX \cup XQ$. The intersection of any two such polygonal lines consists of just the points P and Q; except for those points, they are disjoint. Since there are uncountably many such polygonal lines, there must be one that doesn't meet C at all—which implies that P and Q can be joined by an arc lying entirely in the cocountable set $\mathbb{R}^2 - C$.

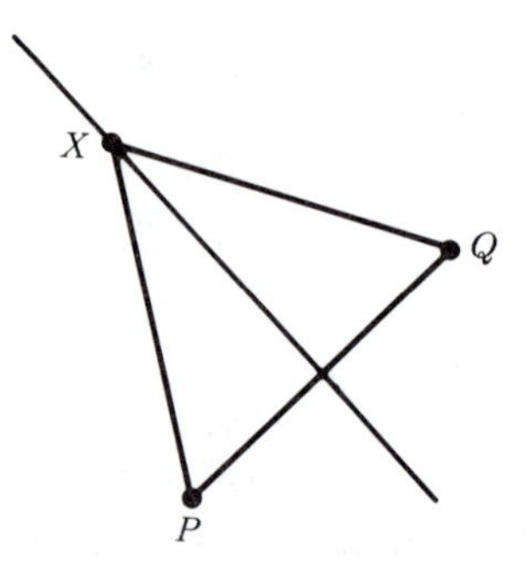

FIGURE 80

Solution 12 B.

12 B

No, a countable subset of the plane with at least two points cannot be connected. One way to prove that is to use the technique of Solution 12 A.

How does one prove that a countable subset C of the line with at least two points cannot be connected? If P and Q are in the set, then find a point A between P and Q that is not in C. Such a point exists because C is countable and there are uncountably many points between P and Q. The parts of C to the right and to the left of A are relatively open subsets of C and constitute a disconnection of C— which shows that C cannot be connected.

A proof for the plane can be made to depend on the theorem for the line. First step: given a countable set C in the plane, find a line that doesn't intersect it. Second step: project C into the line (perpendicularly, or any other pleasant way that comes to mind). Projection is continuous; if the C were connected, its image would be a (countable) connected subset of the line.

Alternative formulation: strengthen the first step a bit so as to produce a line that doesn't intersect C but that has points of C on both sides. The intersections of the countable set with those two sides constitute a disconnection.

Comment. The argument does not prove that countable Hausdorff spaces cannot be connected; weird examples do in fact exist. The present statement concerns only countable subsets of the plane—a shorthand expression that is intended to convey that the topology in question is the one that the set inherits from the ordinary Euclidean topology of the plane.

Solution 12 C.

12 C

The answer is yes: the intersection of a decreasing sequence of compact connected non-empty subsets in the plane must be connected. The plane has not much to do with the answer: any Hausdorff space would do just as well.

Suppose, indeed, that $\{C_n\}$ is a decreasing sequence of compact connected sets with intersection C, and suppose that C is the union of two disjoint closed subsets A and B. Since C_1 (being a compact Hausdorff space) is normal, there exist disjoint open sets U and V in C_1 with $A \subset U$ and $B \subset V$. The compact sets $C_n - (U \cup V)$ form a decreasing sequence whose intersection is $C - (U \cup V)$, which is empty. It follows that $C_m \subset U \cup V$ for some value of m, and hence that C_m is the union of the two disjoint relatively open subsets $C_m \cap U$ and $C_m \cap V$. The connectedness of C_m implies that one of these sets is empty. If, for instance, $C_m \cap U = \varnothing$, then it follows that $C \cap U = \varnothing$, and therefore that $A = \varnothing$. Conclusion: C is connected.

12 D

Solution 12 D.

The answer is no: the plane cannot be a disjoint union of circles. Suppose indeed that it were. Consider any one of the circles and call its diameter d. The circles in the interior of that circle cannot all have diameter greater than $\frac{d}{2}$, for then any two of those interior circles would intersect. Consequence: at least one of those interior circles must have a diameter less than or equal to $\frac{d}{2}$. Apply the same argument, inductively, infinitely often: the result is a decreasing sequence of closed disks with diameters tending to 0. Their intersection doesn't fit into any of the assumed circles.

12 E

Solution 12 E.

The answer is no again: the plane is not a disjoint union of circles, even topological ones. If it were (to save a few syllables, it is convenient for now to call topological circles just simply "circles"), then, by the Schönflies strengthening of the Jordan curve theorem, each one has an interior, and the union of such a circle with its interior (call such a thing a "disk") is homeomorphic to an honest closed disk. Use Zorn's lemma to form a maximal chain of "disks". By compactness, the intersection of such a chain is not empty. If p_0 is a point in that intersection, then by assumption, p_0 belongs to some "circle" C_0. If D is any "disk" in the maximal chain, then $p_0 \in D$, and therefore $C_0 \subset D$. (If there are points of C_0 both inside and outside the "circle" C that is the boundary of D, then the assumed disjointness of the original set of circles would imply that $C \cap C_0 \neq \varnothing$ and therefore that $C = C_0$.) It follows that C_0 is included in the intersection of the chain, and therefore so is D_0. Consider then a point p_0' inside the "circle" C_0. It too lies on some "circle" C_0', included in D_0. The existence of the corresponding "disk" D_0' contradicts the maximality of the original chain.

Solution 12 F.

12 F

The answer is yes: $\mathbb{R}^3$ is a disjoint union of topological circles.

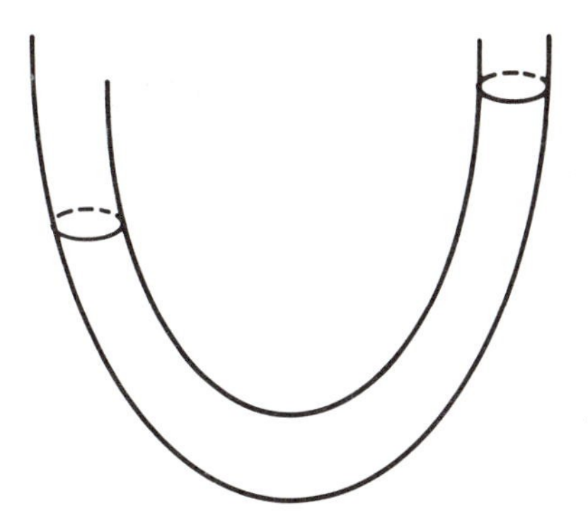

FIGURE 81

What is obvious is that the complement of an infinite open cylinder (such as $\{(x, y, z): x^2 + y^2 < 1\}$) is the union of disjoint (honest) circles. Distort the picture so obtained: replace the cylinder by a U-shaped tube U_1. The set U_1 is homeomorphic to $\mathbb{R}^3$, and the problem reduces to filling it with disjoint (topological) circles. To do so, locate in U_1 a U-

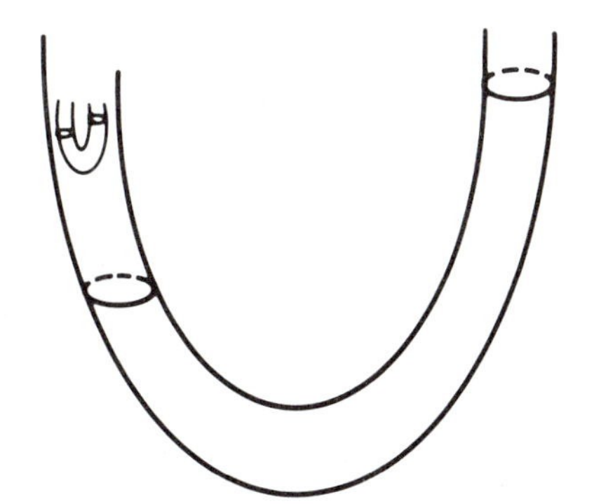

FIGURE 82

shaped open tube U_2, high up in one branch. Fill the complement of U_2 in U_1 with disjoint (topological) circles, and thus reduce the problem to filling U_2. To do so, locate in U_2 a U-shaped open tube U_3 very high up in one branch. Continue ad infinitum, pushing the bent tubes off to infinity.

12 G

Solution 12 G.

In view of the negative answer to the question of space filling honest circles in $\mathbb{R}^2$, most people would probably conjecture that the answer to the question for space filling honest circles in $\mathbb{R}^3$ is also no—it's a surprise to learn that it is yes. The technique requires a little geometric sophistication; here is how it goes.

For each non-negative real number r, let S_r be the sphere

$$\{(x, y, z): x^2 + y^2 + z^2 = r^2\},$$

and note that

$$\bigcup_{r \geq 0} S_r = \mathbb{R}^3.$$

If C is the union of the circles

$$\{(x, y, z): (x - 4k - 1)^2 + y^2 = 1, z = 0\},$$

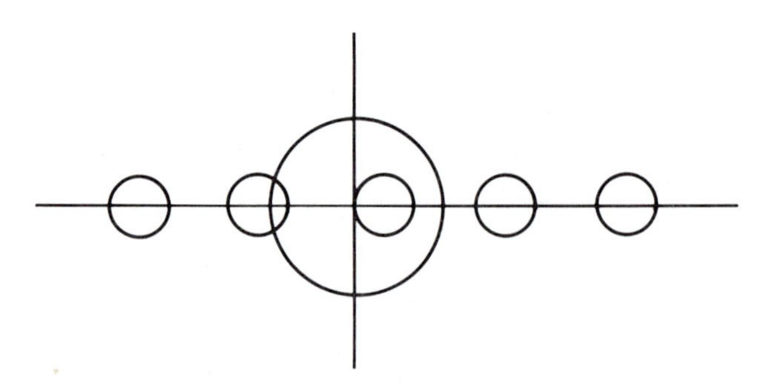

FIGURE **83**

$k = 0, \pm 1, \pm 2, \ldots$, then $C \cap S_r$ consists of exactly two points for each $r > 0$. It follows that

$$\mathbb{R}^3 - C = \bigcup_{r>0} T_r,$$

where

$$T_r = S_r - (C \cap S_r) = S_r \text{ minus two points.}$$

For each $r > 0$ a great circle C_r can be removed from T_r in such a way that $T_r - C_r = T_r' - T_r''$, where T_r' and T_r'' are open hemispheres with one point missing from each. All that remains is to prove that such a punctured open hemisphere can be covered by disjoint "honest" circles. Such a covering can be obtained by, for example, intersecting the hemisphere with a family of planes as indicated by Figure 84.

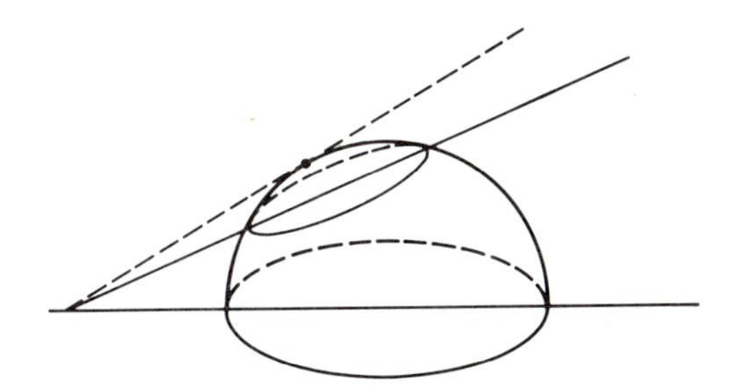

FIGURE 84

Solution 12 H.

12 H

The answer is no: an open interval cannot be a disjoint union of non-degenerate closed intervals.

Suppose, indeed, that an open interval S is such a disjoint union. The main trick in deriving a contradiction from that assumption is to prove that some point of S is simultaneously a limit of the right end points of a sequence of the closed intervals that enter and a limit of the left end points of another such sequence.

To begin the construction, let $I_0 = [a_0, b_0]$ and $J_0 = [c_0, d_0]$ be two of the closed intervals that S is the union of, with the notation chosen so that $b_0 < c_0$. The point x_0 half way between b_0 and c_0 is contained in some interval—call it $I_1 = [a_1, b_1]$ (see Figure 85). The point x_1 half way between b_1 and c_0 is contained in some interval—call it $J_1 = [c_1, d_1]$. Continue similarly, ad infinitum, looking each time at the point half way between the last b and c that were looked at, and baptizing the interval that contains it as the next I or the next J alternately.

FIGURE 85

The half way choice of the x's implies that the even ones converge upward and the odd ones converge downward to the same limit, and hence that the b's converge upward and the c's converge downward to the same limit. The existence of that limit, call it x, leads to a contradiction. Indeed, x must belong to one of the closed intervals that was assumed to cover S. That closed interval either has a part to the left of x, or it has a part to the right of x, or both—in the first case it intersects one of the I's (in fact many of them), in the second case it intersects one of the J's, and in the last case it does both—in any case it contradicts the assumed disjointness.

Comment. The introduction to Problem 12 H commented that the closed-open problem (closed interval as a union of open ones) has a trivially negative solution, and the proof above shows that the solution of the open-closed problem is not *that* trivial but it is also negative. The same proof, with only minor modifications, settles the closed-closed and the open-open problems. The answer is that there is only one way to cover a closed interval with a pairwise disjoint collection of non-degenerate closed intervals, namely by having a singleton collection consisting of the given closed interval alone, and the answer to the open-open question is the same (with "closed" replaced by "open" throughout).

Solution 12 I.

12 I

An attempt to solve the problem directly is likely to lead to confusion; the right thing to do is to sneak up on it. One way to do that is to consider a rectangle of the form in Figure 86. The picture is intended to indicate a rectangle that is almost closed—the only thing that is missing is the top edge. It is obvious, isn't it?, that any such almost closed rectangle is indeed a disjoint union of closed intervals—just use the horizontal ones.

FIGURE 86

The next inspiration is to glue together four such almost closed rectangles that are pairwise disjoint, as indicated by Figure 87, so as to form a rectangle with a hole in the middle. The picture leaves slight gaps between the pieces being glued together, to indicate exactly how they are to be placed; the result after gluing will look like a closed rectangle with an open rectangular hole in the middle as in Figure 88. A better description of a rectangle with a hole in the middle is this: consider two closed rectangles R and S, one inside the other, $R \subset S$, and form the

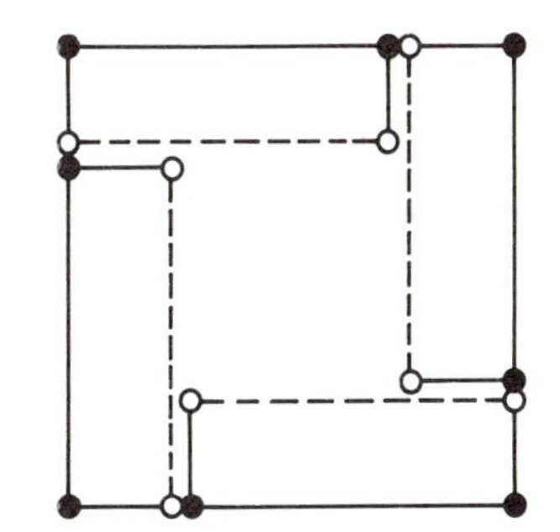

FIGURE 87

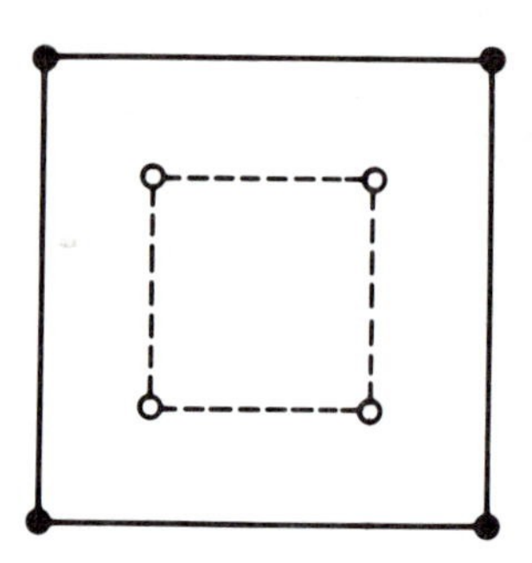

FIGURE 88

difference $S - R$. A possible way to refer to such a difference, such a rectangle with a hole, is to call it an annular rectangle.

Now the inspiration is almost over. Given an arbitrary open rectangle R, write it as the union of an increasing sequence $\{R_0, R_1, R_2, \ldots\}$ of closed rectangles, and then rewrite that union as the disjoint union whose terms are the first rectangle and the successive differences:

$$R = R_0 \cup (R_1 - R_0) \cup (R_2 - R_1) \cup \cdots.$$

The first rectangle R_0 is a disjoint union of closed intervals, and the successive differences are annular rectangles, and, therefore, are also disjoint unions of that kind. Conclusion: R is a disjoint union of closed intervals—and the proof is complete.

12 J Solution 12 J.

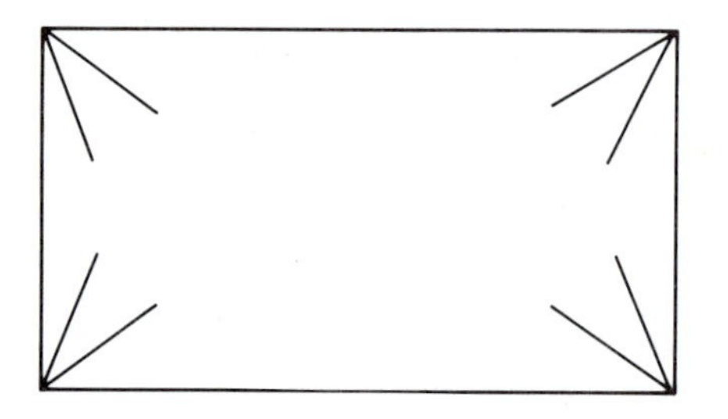

FIGURE 89

Yes, a closed rectangle is a disjoint union of non-empty topological open intervals. As the preliminary discussion indicates, the chief problem is with the corners. To solve that problem, begin by including each of the four corners in a topological open interval, such as, for instance, the V-shaped wedges indicated in Figure 89. That's all it takes; from here on in it's easy. The four edges are open intervals as they stand; all that remains is to cover the part of the interior that the wedges leave empty. To do that, just draw all the horizontal open intervals that join the left vertical edge to the right one, and break them up into subintervals as demanded by the wedges. Near the top and near the bottom, each of the horizontal open intervals becomes five smaller ones, a move a little more toward the center breaks up each of the starting open intervals into three smaller ones, and in the middle they stay whole—no break is necessary.

Solution 12 K.

12 K

The answer is no: the unit interval is not (homeomorphic to) a non-trivial Cartesian product. In this context it doesn't matter whether "interval" is interpreted as closed, or open, or half-open, but the argument is clearest and uses the smallest number of words if we restrict attention to $[0,1]$.

If $[0,1] = X \times Y$, where X and Y are topological spaces, then the factor X is homeomorphic to the section $X \times y$ for each point y in Y; since $X \times y$ is a subset of $X \times Y$, it follows that X is homeomorphic to a subset of $[0,1]$. The same reasoning applies to Y also.

Recall next that a product of two spaces is connected if and only if each factor is connected. Since $[0,1]$ is connected, it follows that X and Y are connected, and hence that X and Y are (homeomorphic to) intervals. In other words: if $[0,1]$ is a Cartesian product at all, then it is the product of two intervals, that is, to a square. It is geometrically and topologically obvious that the interval is not [homeomorphic to] the square.

Isn't it ? Topology offers many high powered ways to distinguish between non-homeomorphic spaces; in the present case a standard low powered one is available. In the interval there exist points (in fact most of them) whose removal makes the space disconnected; in the square there are no such things, and that difference implies that a square and an interval are not homeomorphic.

12 L

Solution 12 L.

Yes, every segment in a convex metric space does have a midpoint, but the result is not elementary.

To get a proof, consider two points a and b in a convex metric space X; assume, as a convenient normalization, that $d(a, b) = 1$. Consider the interval $I = [0, 1]$, and let $\mathbb{F}$ be the set of all isometric mappings from subsets of I into X such that $f(0) = a$ and $f(1) = b$. (A mapping is isometric if it preserves distances: $d(f(a), f(b)) = d(a, b)$ for all a and b in its domain.) The set $\mathbb{F}$ is not empty: it contains, for instance, the function f with the indicated values on the set $\{0, 1\}$.

The set $\mathbb{F}$ is naturally ordered by extension: write $f \subset g$ in case the domain of g includes the domain of f, and f and g are equal on the domain of f. Zorn's lemma applies, and yields a maximal element f of $\mathbb{F}$. Since every isometry is uniformly continuous, the domain of f is closed. (Otherwise, by virtue of uniform continuity, f could be extended, which would contradict its maximality.)

Assertion: dom $f = I$. If not, then $I -$ dom f is a disjoint union of open intervals. If (u, w) is one of them, then, by convexity, there exists a point x in X such that

$$d(f(u), x) + d(x, f(w)) = d(f(u), f(w)) = d(u, w).$$

If v is the point of the interval (u, w) for which

$$d(u, v) = d(f(u), x),$$

then f can be extended to dom $f \cup \{v\}$ as an isometry. Conclusion: the domain of f must be I—and that implies the existence of a midpoint and a lot more.

12 M

Solution 12 M.

Two colors suffice. Indeed: color a region red or blue according as it is in the intersection of an even or an odd number of the interiors of the boundary circles. A boundary crossing (at a point that is not a vertex) changes that number by exactly one.

Comment. The statement and the proof remain valid for straight lines instead of circles.

Chapter 13. Mappings

Solution 13 A.

13 A

The answer is no: consider x^2 and $1 - x$.

For abelian groups the answer is yes. A self-homeomorphism of the interval is a monotone function. If all transformations in the group are increasing, then the end points are simultaneous fixed points. If one of the transformations T is decreasing, then it has a unique fixed point. The commutativity of the group implies that the set of fixed points of any transformation is sent into itself by any other, and hence that the unique fixed point of T is fixed under every element of the group.

Solution 13 B.

13 B

The answer is no: no matter how R' is placed inside R, some point of the country will be represented by the same point in R and R'. The assertion is that of a fixed point theorem, or, to be more precise, fixed point theorems can be used to prove it. Here is how.

To place R' inside R is the same as defining a transformation, call it T, from R into itself with range R'. The statement that R' has on it the same map of the same country reduced in the ration $2 : 1$ implies that if x and y are any two points of R, then $d(Tx, Ty) = \frac{1}{2}d(x, y)$. The conclusion of the pertinent fixed point theorem is that there exists one and only one point x in R such that $Tx = x$, or, in other words, that T has a unique fixed point.

For the proof, observe first that the diameter of R (the maximum distance between two points) is finite. (If, for instance, R is a square of side length 1, then that diameter is equal to $\sqrt{2}$.) It follows that the diameter of the image $T(R)$ $(= R')$ is half as big, and hence that the diameter of the image of the image of R (that is $T^2(R)$) is half as big as that, and so on. The sequence $\{T^n(R)\}$ is a decreasing sequence of closed sets with diameters converging to 0—which implies that the intersection $\bigcap_n T^n(R)$ consists of a single point, say x. Since Tx belongs to every $T^n(R)$, it follows that $Tx = x$. If also $Ty = y$, then $D(x, y) = d(Tx, Ty) \leqq \frac{1}{2}d(x, y)$, and therefore $d(x, y) = 0$—and that's the end of the proof.

Comment. The general statement of the theorem is that if a transformation T of a complete metric X into itself is such that $d(Tx, Ty) \leqq c \cdot d(x, y)$ for some number c in the open unit interval $(0, 1)$ and for all points x and y in X, then T has a unique fixed point. Note that the assumed factor $\frac{1}{2}$ can be replaced by an arbitrary c, and that the assumed equality can be replaced by an inequality, without changing the conclusion. The inequality justifies calling the result the "contraction" mapping theorem, and it does frequently go by that name; it is also called the Banach fixed point theorem. The fixed point that it yields is sometimes described as an "attractive" fixed point, which is a mathematical appellation, not an esthetic one. The reason is that it "attracts" every other point to itself, in this sense: start with a point x_0 in X, and keep forming its iterates $T^n x_0$; the resulting sequence converges to x (no matter which x_0 it started from). The general theorem has wide applications; it can be used, for instance, to prove the so-called Lipschitz existence theorem for solutions of certain differential equations.

Use of the contraction mapping theorem isn't the only way to solve the "map in maps" problem; another solution is to apply the Brouwer fixed point theorem, and still others are possible via ingenious Euclidean geometry.

13 C

Solution 13 C.

The answer is no: the composition of two incompressible transformations can be compressible. Here is one easy example: define S and T on the real line by

$$Sx = -x \qquad \text{and} \qquad Tx = 1 - x.$$

If $SE \subset E$, then $E = SSE \subset SE \subset E$, so that $SE = E$; the transformation S is incompressible. The same argument proves that T is incompressible. The composition TS is easy to compute ($TSx = x + 1$ for every real number x), and turns out to be compressible: it maps the positive real axis, for instance, onto a significantly smaller proper subset of itself.

Comment. Could there be such an example on a compact space? At first blush it might seem that the answer should be no, but, in fact, the example above can easily be converted to a compact one: just add two points, $+\infty$ and $-\infty$, to the real line and extend S and T to them by making each of S and T map each of the added points onto the other.

Alternatively: replace $(-\infty, +\infty)$ by $[-1, +1]$, define S by the same formula as before (reflection about 0), and define T to be a reflection about $\frac{1}{2}$, say, in the sense that it is to map $[\frac{1}{2}, 1]$ linearly onto $[-1, \frac{1}{2}]$ and vice versa. A simple piece of analytic geometry calculation yields the formulas

$$Tx = -3x + 2 \quad \text{when } x \in \left[\frac{1}{2}, 1\right]$$

and

$$Tx = -\frac{x}{3} + \frac{2}{3} \quad \text{when } x \in \left[-1, \frac{1}{2}\right]$$

for T.

Solution 13 D.

13 D

Yes, the concept of topological group does really depend heavily on the connection between the topology and the group structure—and, correspondingly, yes, it is perfectly possible for two topological groups to have the same topological structure and the same group structure, but, nevertheless, not be the same topological group.

One easy example is the set G of all real numbers of the form $r + s\sqrt{2}$ and the set H of all real numbers of the form $r + s\sqrt{3}$, where, in both cases, r and s are rational, and the topology is induced by that of the real line. Trivially G and H are isomorphic as groups; both are isomorphic to the set of all pairs (r, s) of rational numbers. The homeomorphism of the spaces G and H is a standard exercise in elementary topology: any two countable dense sets in the line are homeomorphic. (Establish a homeomorphic correspondence by enumerating both sets and then picking points from them, alternately, so as to mimic the order structure of each in the other.)

To prove that the topological groups G and H are not isomorphic, assume the existence of an isomorphism $f: G \to H$, and reason as follows. By additivity, $f(r) = rf(1)$ for all rational r, and hence, by continuity, $f(\sqrt{2}) = \sqrt{2} \cdot f(1)$, but that cannot be in H.

13 E

Solution 13 E.

Sets, subsets of some fixed set, have a natural group operation: to add two sets, just add their characteristic functions modulo 2, and consider the set that the result is the characteristic function of. In set-theoretic language this addition is the so-called symmetric difference: it is the set of those points that belong to exactly one of the two given sets. With this addition each set is of order two: the sum of a set with itself is empty. The problem can be solved by exhibiting a suitable collection of sets that is a group with respect to symmetric addition, while, at the same time, it is sufficiently closely knit to be connected.

One solution is the collection of all finite unions of left-closed-right-open intervals in the real line (together with the empty set). That is: consider all intervals of the form $[a, b)$, where $-\infty < a < b < +\infty$, and let G be the collection of all sets that are finite unions of such intervals (together with the empty set). Brief meditation ought to convince anyone that G is a group (with respect to addition modulo 2, that is, with respect to symmetric difference); what topology should be used for it?

One well-known and natural topology is a metric one: define the distance between two sets E and F to be $\mu(E + F)$ (where $+$ denotes the present group operation, and where μ is Lebesgue measure). It is plausible enough (and true) that this topology converts G to a topological group. Given two finite unions of half-closed intervals, think about sliding and stretching them and think about splitting and joining them—such thoughts ought to lead to the conclusion that G is connected—and the problem is solved.

13 F

Solution 13 F.

The answer is no: there exists a continuous automorphism on a locally compact abelian group that is not bicontinuous.

To make an example, let C_n be the unit circle in the complex plane for each integer n $(= 0, \pm 1, \pm 2, \ldots)$, and form the Cartesian product $C = \times_n C_n$. The set C, being a direct product of groups, is a group in a natural way; the question is how to topologize it.

Consider the "left half" $\times_{n=-\infty}^{-1} C_n$ and topologize it discreetly; consider the "right half" $\times_{n=0}^{+\infty} C_n$ and topologize it compactly (as the Cartesian product of the compact factors C_n.) Topologize C as the Cartesian product of the left half and the right half; it then becomes, as promised, a locally compact abelian group.

An element of C is a two-way infinite sequence

$$\{x_n : n = 0, \pm 1, \pm 2, \ldots\};$$

let T be the automorphism that sends that sequence onto

$$\{y_n : n = 0, \pm 1, \pm 2, \ldots\},$$

where $y_n = x_{n-1}$ for each n. The inverse image under T of a closed set is always closed (continuity), but the inverse image under T^{-1} of a closed set may fail to be closed. The idea is that pulling compact sets back to discrete sets does no harm, but the other way it does.

Chapter 14. Measures

Solution 14 A.

14 A

It could happen, couldn't it?, that in the judgment of (2) the piece C_1 is the most desirable, with C_2 a close second, whereas C_3 is almost worthless. To be specific, it is possible that

$$\alpha_2(C_1) = \frac{1}{2}, \quad \alpha_2(C_2) = \frac{5}{12}, \quad \text{and} \quad \alpha_2(C_3) = \frac{1}{12}.$$

It could also happen that evaluations of child (3) are always the same as those of (2), so that

$$\alpha_2(C_1) = \frac{1}{2}, \quad \alpha_2(C_2) = \frac{5}{12}, \quad \text{and} \quad \alpha_2(C_3) = \frac{1}{12}.$$

In that case (2) would have said dibs on C_1, and so would (3). When it comes time for (1) to remove one of the pieces C_2 or C_3, he is a free agent—he can choose either one. If either at random or by malice he chooses C_2, then the blemish of the proposed solution becomes visible. In that case, in the judgment of (2), what's left is worth $\frac{1}{2} + \frac{1}{12}$ $(= \frac{7}{12})$, and child (3) agrees—but in order for both of them to be satisfied after the final division they should each receive a piece worth $\frac{4}{12}$, and there is not enough left for that to be possible.

14 B Solution 14 B.

Yes, "you cut, I choose" can be extended to three claimants in a satisfactory manner; here is how.

Let (1) cut a piece of cake that he regards as being exactly one third in value. If (2) thinks it's worth more, he diminishes it to what is a third in his estimation; if he thinks it's worth a third or less, he nods agreeably; and, in either case, having either cut or nodded, he passes the slice, as it now stands to (3). That claimant, (3), judges the slice he is being given, just as (2) did: he diminishes it or blesses it, according to his judgment. If he diminished it, he eats the diminished slice; the other two are convinced that he made a poor bargain (or, at best, a fair one). If he blessed it, but (2) diminished it, then (2) eats it. If both (2) and (3) blessed it (that is, thought that (1) was too modest), then (1) eats it (undiminished, just as he himself had cut it). In any case, one of them eats the slice (with or without it having been diminished), and the remaining two are convinced that there is at least two thirds left for them. They divide what is left by the classical "you cut, I choose" procedure, and the procedure is finished.

Comment. The problem was made popular by the celebrated Polish mathematician Hugo Steinhaus, who was fond of the pretty gems of mathematics, including puzzles and games. He said that during the second world war, in hiding from the invaders of Poland, the cake problem was not just a game—for the group he was with it was a life and death matter of dividing provisions equitably.

The discussion above leaves certain questions unanswered. What happens, for instance, if the number of people involved is not two (the easy case), and not three (where some thought was required), but 100? The answer is that thinking is required only once: the procedure that reduced the case of three to the case of two can be used, inductively, to reduce any n to $n - 1$.

Another problem is introduced by a possible dog-in-the-manger attitude, which could arise even for two people. Suppose that in a "you cut, I choose" procedure, I cut the cake into what I believe to be two exactly equal pieces, but, in your judgment, one of those is worth two thirds of the whole and you choose that one. You and I differ in our evaluations, but we each know what the other thinks, and, in particular, I am aware that you are not only satisfied, but greedily overjoyed. I might be ill-natured about that—I might say that I not only want what's coming to me, but, at the same time, I don't want you to have more than you think you have coming to you. In other words, what I want is to divide the cake C into two pieces C_1 and C_2, so that not only is it true that

$$\alpha_1(C_1) \geqq \frac{1}{2} \quad \text{and} \quad \alpha_2(C_2) \geqq \frac{1}{2},$$

but, in fact, $\alpha_2(C_2) = \frac{1}{2}$. If you are equally selfish, then you want $\alpha_1(C_1) = \frac{1}{2}$. Can these stringent desires be fulfilled?

In mathematical language, what is at issue is an ordered pair (α_1, α_2) of measures, or, equivalently, a single measure whose values are not numbers but 2-dimensional vectors. The problem, in its most general formulation, is to find out which points in the unit square belong to the range of such a measure. What was wanted up to now, and what was achieved, is that the range contains a point in the top right quarter of that square; the new demand is that the range contain the center of the square. The startling and useful insight, Liapounoff's theorem, is the assertion that (for non-atomic measures) the range is always a convex set. That solves the dog-in-the-manger cake problem: since the range contains both $(0,0)$ and $(1,1)$, it must contain the point $(\frac{1}{2}, \frac{1}{2})$ halfway between them. Note, however, that this solution of the dog-in-the-manger cake problem is an existence proof—it gives no hint of a constructive procedure that yields the solution.

(There are many ways to transliterate the name associated with the theorem from the Cyrillic to the Latin alphabet, at least a dozen by a conservative estimate; the one here used is how it appears in the pertinent publication.)

Solution 14 C.

14 C

Here, as often, the right way to begin is backward—assume that there is such a function f and discover what can be said about it.

If f is even mildly regular (continuity is enough for sure) and if $g(x) = \int_1^x f(t)\,dt$, then g is differentiable and $g'(x) = f(x)$ for all x (meaning, of course, for all x in the domain $(1, \infty)$ under consideration). Since $g(x^2) - g(x) = 1$ for all x, it follows that

$$2xg'(x^2) - g'(x) = 0$$

for all x (and, except, for a multiplicative constant, conversely). This equation is equivalent to

$$2xf(x^2) = f(x),$$

so that the problem reduces to solving this functional equation.

Even trial and error is good enough to yield concrete solutions; an easy one is given by

$$f(x) = \frac{1}{\log 2} \cdot \frac{1}{x \log x}.$$

Comment. Is the indefinite integral of this particular f an elementary function? That's a straighforward calculus question, and the answer turns out to be yes; the derivative of $\log \log x$ is exactly $f(x)$.

There are many other continuous solutions. To find them, consider any fundamental domain of the mapping $x \to x^2$ on $(1, \infty)$, for example the interval $[2, 4)$, and on it consider any continuous f such that $f(4-) = \frac{1}{4} f(2)$; the extension to $(1, \infty)$ via the functional equation will be a solution. (Symbols such as $f(4-)$ are popular in certain branches of analysis. They denote limits, so that, for instance, $f(4-)$ is the limit of $f(x)$ as it approaches 4 through values below 4.)

The relation between f and g above is, of course, our old friend, the fundamental theorem of the calculus.

14 D

Solution 14 D.

The answer is no.

Let E be the Cantor set, which has measure zero, and write $\varepsilon_n = \frac{1}{3^n}$. Assertion: if $\mu(I_n) < \varepsilon_n$, then $E \not\subset \bigcup_n I_n$. Indeed: since $\mu(I_1) < \varepsilon_1$, the set I_1 is disjoint from either the right or the left half of E; let E_1 be that half. Since $\mu(I_2) < \varepsilon_2$, the set I_2 is disjoint from one of the halves, say E_2, of E_1. Continue this way inductively. Consequence: $\{E_n\}$ is a decreasing sequence of compact sets; the intersection $\bigcap_n E_n$ is a non-empty subset of E and is disjoint from $\bigcup_n I_n$.

14 E

Solution 14 E.

The answer is yes: there exists a measurable function on $\mathbb{R}^2$ that is essentially unbounded on every measurable rectangle of positive measure.

Let $r_1, r_2, r_3, \ldots$ be an enumeration of the rational numbers, and let I_n be the open interval with center r_n and radius $\frac{1}{2^n}$, $n = 1, 2, 3, \ldots$. An elementary measure-theoretic argument shows that the real numbers that belong to I_n for infinitely many values of n form a set of measure zero. (The clue is that the series $\sum_n \frac{1}{2^n}$ converges. The probabilistically sophisticated reader might recall the Borel–Cantelli lemma.) It follows that if χ_n is the characteristic function of I_n, then the sequence $\{\chi_1(x), \chi_2(x), \chi_3(x), \ldots\}$ is finitely non-zero for almost every x, which implies, of course, that the series

$$\sum_{n=1}^{\infty} \frac{\chi_n(x)}{|x - r_n|}$$

converges at almost every x. The sum defines an almost-everywhere finite-valued measurable function that is essentially unbounded on every non-empty open set. [Note: the construction could have been carried out with $\frac{1}{x^2}$ in place of $\frac{1}{x}$, that is with

$$\sum_{n=1}^{\infty} \frac{\chi_n(x)}{|x - r_n|^2}$$

in place of

$$\sum_{n=1}^{\infty} \frac{\chi_n(x)}{|x - r_n|},$$

or with $\frac{1}{\sqrt{x}}$, or, in fact, with any function that is essentially unbounded near 0.] Write

$$k(x, y) = \sum_{n=1}^{\infty} \frac{\chi_n(x - y)}{|x - r_n|^2};$$

the proof will be completed by showing that k is essentially unbounded on every measurable rectangle of positive measure.

Fix a (large) positive number M, and, for each $n = 1, 2, 3, \ldots$, consider the points (x, y) for which

$$\frac{\chi_n(x - y)}{|x - y - r_n|} > M.$$

These are the points (x, y) for which $x - y \in I_n \cap J_n$, where J_n is the open interval with center r_n and radius $\frac{1}{M}$. If, that is, δ is the mapping from $\mathbb{R}^2$ to $\mathbb{R}$ defined by

$$\delta(x, y) = x - y,$$

and if D is the (dense) open set $\bigcup_n (I_n \cap J_n)$, then

$$k^{-1}((M, \infty)) \supset \bigcup_n \{(x, y): \frac{\chi_n(x - y)}{|x - y - r_n|} > M\} = \delta^{-1}(D).$$

To finish the proof it is therefore sufficient to prove that, for each M, the set $\delta^{-1}(D)$ meets every measurable rectangle of positive measure in a set of positive measure.

Suppose that $E \times F$ is a measurable rectangle of positive measure. Assertion: $D \cap (E - F)$ has positive measure. To prove the assertion, note that there exists a set G of positive measure, and there exist real numbers α and β, such that $G + \alpha \subset E$ and $G + \beta \subset F$. It follows that

$$E - F \supset (G - G) + (\alpha - \beta),$$

and hence that $E - F$ includes a non-empty open set. This, in turn, implies that $D \cap (E - F)$ includes a set of the same kind, and hence is a set of positive measure.

The proof will be completed by showing that

$$\lambda(\delta^{-1}(D) \cap (E \times F)) > 0,$$

where $\lambda = \mu \times \mu$ is planar Lebesgue measure, the Cartesian square of the linear Lebesgue measure μ. For each real number y,

$$\begin{aligned}(\delta^{-1}(D) \cap (E \times F))^y &= \{x: (x, y) \in \delta^{-1}(D) \cap (E \times F)\} \\ &= \{x: (x, y) \in \delta^{-1}(D)\} \cap \{x: (x, y) \in E \times F\} \\ &= (D + y) \cap E.\end{aligned}$$

Since, therefore,

$$\begin{aligned}\lambda(\delta^{-1}(D) \cap (E \times F)) &= \int \mu((\delta^{-1}(D) \cap (E \times F))^y)\, d\mu(y) \\ &= \int \mu((D + y) \cap E)\, d\mu(y),\end{aligned}$$

the conclusion follows from "the average theorem" of Lebesgue integration; according to that theorem the value of the last integral is, in fact, equal to $\mu(D)\mu(E)$. The proof is complete.

Comment. It's a corollary of what was just proved that there exists a planar set of positive measure that includes no measurable rectangle of positive measure. To see that, consider the function k constructed above, and note that for M sufficiently large the set $\{(x, y): k(x, y) < M\}$ has positive measure.

Solution 14 F.

14 F

The answer is yes.

Suppose that μ is Lebesgue measure and ν is another measure on the Borel sets of the unit interval such that $\nu(E) = \frac{1}{2}$ whenever $\mu(E) = \frac{1}{2}$. For each (temporarily fixed) positive integer n, let $E(1), \ldots, E(2n)$ be the dyadic intervals of length $\frac{1}{2^n}$:

$$E(1) = \left[0, \frac{1}{2^n}\right], E(2) = \left[\frac{1}{2^n}, \frac{2}{2^n}\right], \ldots, E(2n) = \left[\frac{2^n - 1}{2^n}, 1\right].$$

Assertion:

$$\nu(E(i)) = \nu(E(j)) = \frac{1}{2^n}$$

for each i and j $(= 1, \ldots, 2^n)$. The trick for proving the assertion is to use $E(i)$ and $E(j)$ to form a couple of clever sets of Lebesgue measure $\frac{1}{2}$, and the way to do *that* is to glue together almost half of the dyadic intervals of length $\frac{1}{2^n}$. To be specific: find $2^{n-1} - 1$ of the $E(k)$'s all distinct from one another and from both $E(i)$ and $E(j)$. The union of those $E(k)$'s with $E(i)$ is a set of Lebesgue measure $\frac{1}{2}$, and therefore, by the assumption on ν, it has ν measure $\frac{1}{2}$, and exactly the same is true of the union of the $E(k)$'s and $E(j)$. The assertion about $\nu(E(i))$ and $\nu(E(j))$ is an immediate consequence.

In other words: any two dyadic intervals of the same length have the same ν measure, namely their Lebesgue measure. Conclusion: μ and ν agree on the algebra of all dyadic sets, and, therefore, they agree on all Borel sets.

Comment. The number $\frac{1}{2}$ plays no essential role here; the theorem remains true if $\frac{1}{2}$ is replaced by any number c, $0 < c < 1$.

14 G

Solution 14 G.

The answer is no: one suitable counterexample is the collection known to students of probability as the Rademacher sets. To see them, write

$$E_1 = \left[0, \frac{1}{2}\right],$$

$$E_2 = \left[0, \frac{1}{4}\right] \cup \left[\frac{1}{2}, \frac{3}{4}\right],$$

and, in general,

$$E_n = \left[0, \frac{1}{2^n}\right] \cup \left[\frac{2}{2^n}, \frac{3}{2^n}\right] \cup \cdots \cup \left[\frac{2^n - 2}{2^n}, \frac{2^n - 1}{2^n}\right].$$

An alternative description of these Rademacher sets goes as follows: E_n is the set of those points x in $[0, 1]$ in whose dyadic expansion the nth digit is 0. In this description the dyadically rational numbers (that is, the numbers of the form $\frac{k}{2^n}$) make for some annoying but unimportant ambiguity; it's unimportant because they constitute a countable set, and hence a set of measure 0. Note, by the way, that

$$\mu(E_n) = \frac{1}{2}$$

for $n = 1, 2, 3, \ldots$. (Here μ is Lebesgue measure, of course.)

The sequence $\{E_1, E_2, E_3, \ldots\}$ is stochastically independent, meaning that the measure of the intersection of any k terms, $k = 1, 2, 3, \ldots$, is equal to the product of their measures (and hence equal to $\frac{1}{2^k}$). Consequence: the intersection of every infinite collection of E's has measure 0, and that concludes the proof of the negative answer.

14 H

Solution 14 H.

The answer is yes: uncountability is enough to imply the existence of a large intersection. In view of the comment following the statement of the problem, there is no loss of generality in assuming, in the course of the proof, that the given collection $\mathfrak{C}$ of measurable sets of positive measure consists of sets whose measures are bounded away from 0.

Recall now that the measure algebra consisting of the equivalence classes of measurable sets modulo sets of measure 0 is a separable metric space, with the distance between two sets A and B being defined by the measure $\mu(A + B)$ of the symmetric difference $A + B$. (The symmetric difference of two sets is the set whose characteristic function is obtained from the given ones by addition modulo 2.) By separability, the collection $\mathfrak{C}$ has a condensation point in that space; that is, there exists a set P of positive measure such that each ball with center P has an uncountable intersection with $\mathfrak{C}$. All that really matters is that P is a cluster point of $\mathfrak{C}$. For each $n = 1, 2, 3, \ldots$, choose A_n in $\mathfrak{C}$ so that

$$\mu(P + A_n) < \frac{1}{2^{n+1}}\mu(P).$$

It follows that

$$\begin{aligned}\mu\left(\bigcap_{n \geqq 1} A_n\right) &\geqq \mu\left(\bigcap_{n \geqq 1}(P \cap A_n)\right)\\ &= \mu(P) - \mu\left(\bigcup_{n \geqq 1}(P - A_n)\right)\\ &\geqq \mu(P) - \sum_{n \geqq 1}\mu(P + A_n)\\ &> \mu(P) - \sum_{n \geqq 1}\frac{1}{2^{n+1}}\mu(P) = \frac{1}{2}\mu(P),\end{aligned}$$

and that implies the desired conclusion.

Solution 14 I. 14 I

The answer is yes. To see the proof, observe first of all that there is no loss of generality in assuming that the prescribed collection of sets is countably infinite, say $\{E_1, E_2, E_3, \ldots\}$. The question is whether there are any points that belong to infinitely many of the E's, or, in known, classical, terms, whether the lim sup of the sequence $\{E_1, E_2, E_3, \ldots\}$, call it E^*, is non-empty. Since

$$E^* = \bigcap_{n \geqq 1}\bigcup_{k \geqq n} E_k,$$

and since the sequence $\{\bigcup_{k \geqq n} E_k : k = 1, 2, 3, \ldots\}$ of partial unions is decreasing, it follows that if $\mu(E_n) \geqq \varepsilon > 0$ for all n, then

$$\mu(E^*) = \lim_{n \to \infty} \mu(\bigcup_{k \geqq n} E_k) \geqq \liminf_{n \to \infty} \mu(E_n) \geqq \varepsilon.$$

Consequence: the lim sup, having positive measure, is certainly not empty.

Comment. Caution: if a sequence has an infinite subsequence with a large intersection (in any sense of the word) then the lim sup of the sequence is large (in that sense), but not conversely. Consider, for example, the sequence $\{E_1', E_2', E_3', \ldots\}$ of the complements of the Rademacher sets. That complementary sequence has exactly the same measure-theoretic properties as the Rademacher sequence itself, and, in particular, it has no infinite subsequence with positive intersection. To get a new piece of information look at the lim inf, call it E_*, of the original sequence $\{E_1, E_2, E_3, \ldots\}$. Since

$$E_* = \liminf_{n \to \infty} E_n = \bigcup_{n \geqq 1} \bigcap_{k \geqq n} E_k,$$

and since

$$\mu\left(\bigcap_{k \geqq n} E_k\right) = 0$$

for all n, it follows that

$$\mu(E_*) = \lim_{n \to \infty} \mu\left(\bigcap_{k \geqq n} E_k\right) = 0,$$

and hence that the lim sup E'^* of the complements is full (that is, has measure 1). In other words, the complementary sequence has a large lim sup but does not have a subsequence with large intersection.